An Introduction to
Seismology

An Introduction to Seismology

Edited by
Jonathan Frost

Larsen & Keller
www.larsen-keller.com

An Introduction to Seismology
Edited by Jonathan Frost
ISBN: 978-1-63549-258-3 (Hardback)

📚 Larsen & Keller

Published by Larsen and Keller Education,
5 Penn Plaza,
19th Floor,
New York, NY 10001, USA

Cataloging-in-Publication Data

An introduction to seismology / edited by Jonathan Frost.
    p. cm.
Includes bibliographical references and index.
ISBN 978-1-63549-258-3
1. Seismology. 2. Earthquakes. 3. Geophysics.
I. Frost, Jonathan.
QE534.3 .I58 2017
551.22--dc23

The publisher's policy is to use permanent paper from mills that operate a sustainable forestry policy. Furthermore, the publisher ensures that the text paper and cover boards used have met acceptable environmental accreditation standards.

Printed and bound in the United States of America.

For more information regarding Larsen and Keller Education and its products, please visit the publisher's website www.larsen-keller.com

Table of Contents

Preface

This book provides comprehensive insights into the field of seismology. It talks in detail about the different concepts and scientific theories related to this field. Seismology refers to the study of the earthquakes and their causes. It also includes the study of the effects of earthquakes and other seismically caused disasters like volcanoes, tectonic explosions, tsunamis, etc. The book is compiled in such a manner, that it will provide in-depth knowledge about the latest theory about earthquakes. The topics included in it are of utmost significance and are bound to provide incredible insights to readers. Those in search of information to further their knowledge will be greatly assisted by this textbook.

To facilitate a deeper understanding of the contents of this book a short introduction of every chapter is written below:

Chapter 1- The study of earthquakes is known as seismology. It also studies the effect earthquakes have on the environment. A person who studies seismology is known as a seismologist. This text will provide an integrated understanding of seismology.

Chapter 2- This section explains to the reader the important key concepts of seismology. Seismic waves are waves that travel through Earth's layers. Seismic waves can be characterized into certain types, such as P-wave, S-wave, Rayleigh wave and Love wave. Some of the concepts explained in this section are seismic waves, ductility of rocks, seismic, hotspot and mantle plume.

Chapter 3- Reflection seismology is the method that uses the principles of seismology to assess the substratum after a reflected seismic wave. Seismic migration is the transferring of seismic events because of the complexity in the initial geographical area. The following text provides the reader with an in-depth understanding of reflection seismology.

Chapter 4- The structure of the Earth is spherical and has an inner core and an outer core. The crust of the Earth is the outer solid shell which is generated by igneous processes. The chapter strategically encompasses and incorporates the basic understanding of the structure of the Earth.

Chapter 5- An earthquake is the shaking of the surface of the Earth. Earthquakes can severely damage life and property and can also result in decline in the economy. It can cause damage that can take years to recover from, and they can be measured by using observations from seismometers. Some of the aspects elucidated within this section are aftershock, foreshock, induced seismicity, cryoseism and submarine earthquake.

Chapter 6- The extent of damage earthquakes can cause is very extreme. Governments take steps in order to analyze and then to take precautionary steps regarding them. Some of these are earthquake preparedness, earthquake warning system, seismic analysis and seismic retrofit. Earthquakes can best be understood in confluence with the major topics listed in the following chapter.

Chapter 7- Earthquake engineering is a branch of engineering that studies buildings and structures in order to prevent them from earthquakes. The basic aim of this subject is to make buildings resistant to earthquakes. This text helps the reader in developing an in-depth understanding of the subject matter.

Finally, I would like to thank the entire team involved in the inception of this book for their valuable time and contribution. This book would not have been possible without their efforts. I would also like to thank my friends and family for their constant support.

Editor

Introduction to Seismology

The study of earthquakes is known as seismology. It also studies the effect earthquakes have on the environment. A person who studies seismology is known as a seismologist. This text will provide an integrated understanding of seismology.

Seismology is the scientific study of earthquakes and the propagation of elastic waves through the Earth or through other planet-like bodies. The field also includes studies of earthquake environmental effects, such as tsunamis as well as diverse seismic sources such as volcanic, tectonic, oceanic, atmospheric, and artificial processes (such as explosions). A related field that uses geology to infer information regarding past earthquakes is paleoseismology. A recording of earth motion as a function of time is called a seismogram. A seismologist is a scientist who does research in seismology.

History

Scholarly interest in earthquakes can be traced back to antiquity. Early speculations on the natural causes of earthquakes were included in the writings of Thales of Miletus (c. 585 BCE), Anaximenes of Miletus (c. 550 BCE), Aristotle (c. 340 BCE) and Zhang Heng (132 CE).

In 132 CE, Zhang Heng of China's Han dynasty designed the first known seismoscope.

In 1664, Athanasius Kircher argued that earthquakes were caused by the movement of fire within a system of channels inside the Earth.

In 1703, Martin Lister (1638 to 1712) and Nicolas Lemery (1645 to 1715) proposed that earthquakes were caused by chemical explosions within the earth.

The Lisbon earthquake of 1755, coinciding with the general flowering of science in Europe, set in motion intensified scientific attempts to understand the behaviour and causation of earthquakes. The earliest responses include work by John Bevis (1757) and John Michell (1761). Michell determined that earthquakes originate within the Earth and were waves of movement caused by "shifting masses of rock miles below the surface."

From 1857, Robert Mallet laid the foundation of instrumental seismology and carried out seismological experiments using explosives. He is also responsible for coining the word "seismology".

In 1897, Emil Wiechert's theoretical calculations led him to conclude that the Earth's interior consists of a mantle of silicates, surrounding a core of iron.

In 1906 Richard Dixon Oldham identified the separate arrival of P-waves, S-waves and surface waves on seismograms and found the first clear evidence that the Earth has a central core.

In 1910, after studying the 1906 San Francisco earthquake, Harry Fielding Reid put forward the "elastic rebound theory" which remains the foundation for modern tectonic studies. The development of this theory depended on the considerable progress of earlier independent streams of work on the behaviour of elastic materials and in mathematics.

In 1926, Harold Jeffreys was the first to claim, based on his study of earthquake waves, that below the mantle, the core of the Earth is liquid.

In 1937, Inge Lehmann determined that within the earth's liquid outer core there is a solid *inner* core.

By the 1960s, earth science had developed to the point where a comprehensive theory of the causation of seismic events had come together in the now well-established theory of plate tectonics.

Types of Seismic Wave

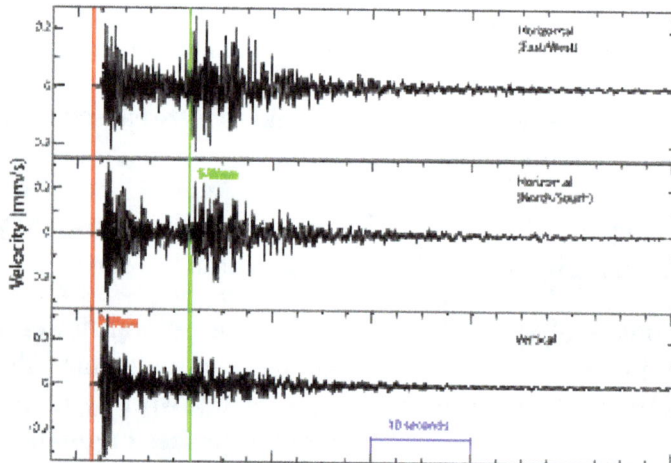

Seismogram records showing the three components of ground motion. The red line marks the first arrival of P-waves; the green line, the later arrival of S-waves.

Seismic waves are elastic waves that propagate in solid or fluid materials. They can be divided into *body waves* that travel through the interior of the materials; *surface waves* that travel along surfaces or interfaces between materials; and *normal modes*, a form of standing wave.

Body Waves

There are two types of body waves, Pressure waves or Primary waves (P-waves) and Shear or Secondary waves (S-waves). P-waves, are longitudinal waves that involve compression and expansion in the direction that the wave is moving. P-waves are the fastest waves in solids and are therefore the first waves to appear on a seismogram. S-waves are transverse waves that move perpendicular to the direction of propagation. S-waves are slower than P-waves. Therefore, they appear later than P-waves on a seismogram. Fluids cannot support perpendicular motion, so S-waves only travel in solids.

Surface Waves

The two main surface wave types are Rayleigh waves, which have some compressional motion, and Love waves, which do not. Rayleigh waves result from the interaction of vertically polarized P- and S-waves that satisfy the boundary conditions on the surface. Love waves can exist in the presence of a subsurface layer, and are only formed by horizontally polarized S-waves. Surface waves travel more slowly than P-waves and S-waves; however, because they are guided by the Earth's surface and their energy is thus trapped near the surface, they can be much stronger than body waves, and can be the largest signals on earthquake seismograms. Surface waves are strongly excited when their source is close to the surface, as in a shallow earthquake or a near surface explosion.

Normal Modes

Both body and surface waves are traveling waves; however, large earthquakes can also make the Earth "ring" like a bell. This ringing is a mixture of normal modes with discrete frequencies and periods of an hour or shorter. Motion caused by a large earthquake can be observed for up to a month after the event. The first observations of normal modes were made in the 1960s as the advent of higher fidelity instruments coincided with two of the largest earthquakes of the 20th century - the 1960 Valdivia earthquake and the 1964 Alaska earthquake. Since then, the normal modes of the Earth have given us some of the strongest constraints on the deep structure of the Earth.

Earthquakes

One of the first attempts at the scientific study of earthquakes followed the 1755 Lisbon earthquake. Other notable earthquakes that spurred major advancements in the science of seismology include the 1857 Basilicata earthquake, 1906 San Francisco earthquake, the 1964 Alaska earthquake, the 2004 Sumatra-Andaman earthquake, and the 2011 Great East Japan earthquake.

Controlled Seismic Sources

Seismic waves produced by explosions or vibrating controlled sources are one of the primary methods of underground exploration in geophysics (in addition to many different electromagnetic methods such as induced polarization and magnetotellurics). Controlled-source seismology has been used to map salt domes, anticlines and other geologic traps in petroleum-bearing rocks, faults, rock types, and long-buried giant meteor craters. For example, the Chicxulub Crater, which was caused by an impact that has been implicated in the extinction of the dinosaurs, was localized to Central America by analyzing ejecta in the Cretaceous–Paleogene boundary, and then physically proven to exist using seismic maps from oil exploration.

Detection of Seismic Waves

Installation for a temporary seismic station, north Iceland highland.

Seismometers are sensors that sense and record the motion of the Earth arising from elastic waves. Seismometers may be deployed at the Earth's surface, in shallow vaults, in boreholes, or underwater. A complete instrument package that records seismic signals is called a seismograph. Networks of seismographs continuously record ground motions around the world to facilitate the monitoring and analysis of global earthquakes and other sources of seismic activity. Rapid location of earthquakes makes tsunami warnings possible because seismic waves travel considerably faster than tsunami waves. Seismometers also record signals from non-earthquake sources ranging from explosions (nuclear and chemical), to local noise from wind or anthropogenic activities, to incessant signals generated at the ocean floor and coasts induced by ocean waves (the global microseism), to cryospheric events associated with large icebergs and glaciers. Above-ocean meteor strikes with energies as high as 4.2×10^{13} J (equivalent to that released by an explosion of ten kilotons of TNT) have been recorded by seismographs, as have a number of industrial accidents and terrorist bombs and events (a field of study referred to as forensic seismology). A major long-term motivation for the global seismographic monitoring has been for the detection and study of nuclear testing.

Mapping the Earth's Interior

Seismic velocities and boundaries in the interior of the Earth sampled by seismic waves

Because seismic waves commonly propagate efficiently as they interact with the internal structure of the Earth, they provide high-resolution noninvasive methods for studying the planet's interior. One of the earliest important discoveries (suggested by Richard Dixon Oldham in 1906 and definitively shown by Harold Jeffreys in 1926) was that the outer core of the earth is liquid. Since S-waves do not pass through liquids, the liquid core causes a "shadow" on the side of the planet opposite of the earthquake where no direct S-waves are observed. In addition, P-waves travel much slower through the outer core than the mantle.

Processing readings from many seismometers using seismic tomography, seismologists have mapped the mantle of the earth to a resolution of several hundred kilometers. This has enabled scientists to identify convection cells and other large-scale features such as the large low-shear-velocity provinces near the core–mantle boundary.

Seismology and Society

Earthquake Prediction

Forecasting a probable timing, location, magnitude and other important features of a forthcoming seismic event is called earthquake prediction. Various attempts have been made by seismologists and others to create effective systems for precise earthquake predictions, including the VAN method. Most seismologists do not believe that a system to provide timely warnings for individual earthquakes has yet been developed, and many believe that such a system would be unlikely to give useful warning of impending seismic events. However, more general forecasts routinely predict seismic hazard. Such forecasts estimate the probability of an earthquake of a particular size affecting a particular location within a particular time-span, and they are routinely used in earthquake engineering.

Public controversy over earthquake prediction erupted after Italian authorities indicted six seismologists and one government official for manslaughter in connection with a magnitude 6.3 earthquake in L'Aquila, Italy on April 5, 2009. The indictment has been widely perceived as an indictment for failing to predict the earthquake and has drawn condemnation from the American Association for the Advancement of Science and the American Geophysical Union. The indictment claims that, at a special meeting in L'Aquila the week before the earthquake occurred, scientists and officials were more interested in pacifying the population than providing adequate information about earthquake risk and preparedness.

Engineering Seismology

Engineering seismology is the study and application of seismology for engineering purposes. It generally applied to the branch of seismology that deals with the assessment of the seismic hazard of a site or region for the purposes of earthquake engineering. It is, therefore, a link between earth science and civil engineering. There are two principal components of engineering seismology. Firstly, studying earthquake history (e.g. historical and instrumental catalogs of seismicity) and tectonics to assess the earthquakes that could occur in a region and their characteristics and frequency of occurrence. Secondly, studying strong ground motions generated by earthquakes to assess the expected shaking from future earthquakes with similar characteristics. These strong ground motions could either be observations from accelerometers or seismometers or those simulated by computers using various techniques.

References

- Shearer, Peter M. (2009). Introduction to Seismology (Second ed.). Cambridge University Press. ISBN 978-0-521-70842-5.

- Stein, Seth; Wysession, Michael (2002). An Introduction to Seismology, Earthquakes and Earth Structure. Wiley-Blackwell. ISBN 978-0-86542-078-6.

- Gubbins, David (1990). Seismology and Plate Tectonics. Cambridge University Press. ISBN 0-521-37141-4.

- Hall, Stephen S. (2011). "Scientists on trial: At fault?". Nature. 477 (7364): 264–269. Bibcode:2011Natur.477..264H. doi:10.1038/477264a. PMID 21921895.

Key Concepts of Seismology

This section explains to the reader the important key concepts of seismology. Seismic waves are waves that travel through Earth's layers. Seismic waves can be characterized into certain types, such as P-wave, S-wave, Rayleigh wave and Love wave. Some of the concepts explained in this section are seismic waves, ductility of rocks, seismic, hotspot and mantle plume.

Seismic Wave

Seismic waves are waves of energy that travel through the Earth's layers, and are a result of earthquakes, volcanic eruptions, magma movement, large landslides and large man-made explosions that give out low-frequency acoustic energy. Many other natural and anthropogenic sources create low-amplitude waves commonly referred to as ambient vibrations. Seismic waves are studied by geophysicists called seismologists. Seismic wave fields are recorded by a seismometer, hydrophone (in water), or accelerometer.

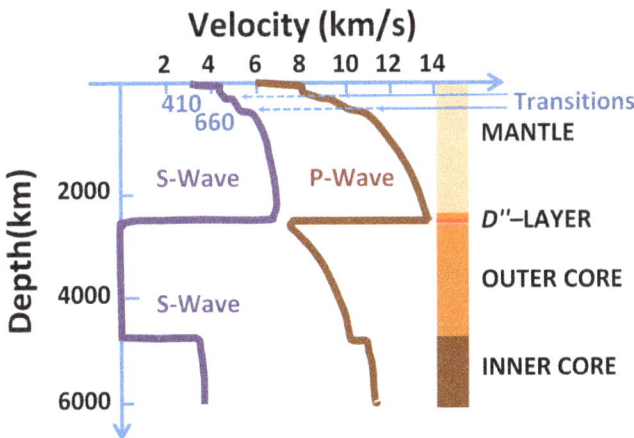

Velocity of seismic waves in the Earth versus depth. The negligible S-wave velocity in the outer core occurs because it is liquid, while in the solid inner core the S-wave velocity is non-zero.

The propagation velocity of the waves depends on density and elasticity of the medium. Velocity tends to increase with depth and ranges from approximately 2 to 8 km/s in the Earth's crust, up to 13 km/s in the deep mantle.

Earthquakes create distinct types of waves with different velocities; when reaching seismic observatories, their different travel times help scientists to locate the source of the hypocenter. In geophysics the refraction or reflection of seismic waves is used for research into the structure of the Earth's interior, and man-made vibrations are often generated to investigate shallow, subsurface structures.

Types

Among the many types of seismic waves, one can make a broad distinction between *body waves*, which travel through the Earth, and *surface waves*, which travel at the Earth's surface.

Other modes of wave propagation exist than those described in this article; though of comparatively minor importance for earth-borne waves, they are important in the case of asteroseismology.

- Body waves travel through the interior of the Earth.

- Surface waves travel across the surface. Surface waves decay more slowly with distance than do body waves, which travel in three dimensions.

- Particle motion of surface waves is larger than that of body waves, so surface waves tend to cause more damage.

Body Waves

Body waves travel through the interior of the Earth along paths controlled by the material properties in terms of density and modulus (stiffness). The density and modulus, in turn, vary according to temperature, composition, and material phase. This effect resembles the refraction of light waves. Two types of particle motion result in two types of body waves: *Primary* and *Secondary* waves.

Primary Waves

Primary waves (P-waves) are compressional waves that are longitudinal in nature. P waves are pressure waves that travel faster than other waves through the earth to arrive at seismograph stations first, hence the name "Primary". These waves can travel through any type of material, including fluids, and can travel at nearly twice the speed of S waves. In air, they take the form of sound waves, hence they travel at the speed of sound. Typical speeds are 330 m/s in air, 1450 m/s in water and about 5000 m/s in granite.

Secondary Waves

Secondary waves (S-waves) are shear waves that are transverse in nature. Following an earthquake event, S-waves arrive at seismograph stations after the faster-moving P-waves and displace the ground perpendicular to the direction of propagation. Depending on the propagational direction, the wave can take on different surface characteristics; for example, in the case of horizontally polarized S waves, the ground moves alternately to one side and then the other. S-waves can travel only through solids, as fluids (liquids and gases) do not support shear stresses. S-waves are slower than P-waves, and speeds are typically around 60% of that of P-waves in any given material.

Surface Waves

Seismic surface waves travel along the Earth's surface. They can be classified as a form of mechanical surface waves. They are called surface waves, as they diminish as they get further from the surface. They travel more slowly than seismic body waves (P and S). In large earthquakes, surface waves can have an amplitude of several centimeters.

Rayleigh Waves

Rayleigh waves, also called ground roll, are surface waves that travel as ripples with motions that are similar to those of waves on the surface of water (note, however, that the associated particle motion at shallow depths is retrograde, and that the restoring force in Rayleigh and in other seismic waves is elastic, not gravitational as for water waves). The existence of these waves was predicted by John William Strutt, Lord Rayleigh, in 1885. They are slower than body waves, roughly 90% of the velocity of S waves for typical homogeneous elastic media. In the layered medium (like the crust and upper mantle) the velocity of the Rayleigh waves depends on their frequency and wavelength.

Love Waves

Love waves are horizontally polarized shear waves (SH waves), existing only in the presence of a semi-infinite medium overlain by an upper layer of finite thickness. They are named after A.E.H. Love, a British mathematician who created a mathematical model of the waves in 1911. They usually travel slightly faster than Rayleigh waves, about 90% of the S wave velocity, and have the largest amplitude.

Stoneley Waves

A Stoneley wave is a type of boundary wave (or interface wave) that propagates along a solid-fluid boundary or, under specific conditions, also along a solid-solid boundary. Amplitudes of Stoneley waves have their maximum values at the boundary between the two contacting media and decay exponentially towards the depth of each of them. These waves can be generated along the walls of a fluid-filled borehole, being an important source of coherent noise in VSPs and making up the low frequency component of the source in sonic logging. The equation for Stoneley waves was first given by Dr. Robert Stoneley (1894 - 1976), Emeritus Professor of Seismology, Cambridge.

Free Oscillations of the Earth

Free oscillations of the Earth are standing waves, the result of interference between two surface waves traveling in opposite directions. Interference of Rayleigh waves results in *spheroidal oscillation S* while interference of Love waves gives *toroidal oscillation T*. The modes of oscillations are specified by three numbers, e.g., $_nS_l^m$, where l is the angular order number. The number m is the azimuthal order number. It may take on $2l+1$ values from $-l$ to $+l$. The number n is the *radial order number*. It means the wave with n zero crossings in radius. For spherically symmetric Earth the period for given n and l does not depend on m.

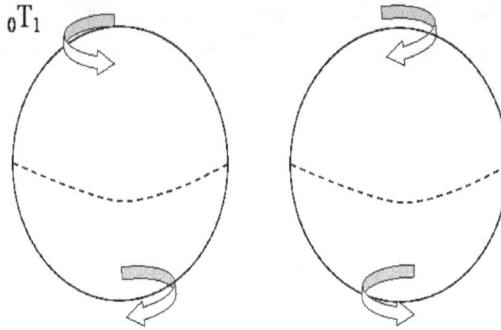

The sense of motion for toroidal $_0T_1$ oscillation for two moments of time.

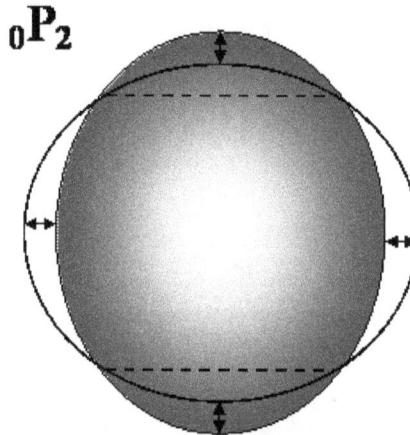

The scheme of motion for spheroidal $_0S_2$ oscillation.Dashed lines give nodal (zero) lines. Arrows give the sense of motion.

Some examples of spheroidal oscillations are the "breathing" mode $_0S_0$, which involves an expansion and contraction of the whole Earth, and has a period of about 20 minutes; and the "rugby" mode $_0S_2$, which involves expansions along two alternating directions, and has a period of about 54 minutes. The mode $_0S_1$ does not exist because it would require a change in the center of gravity, which would require an external force.

Of the fundamental toroidal modes, $_0T_1$ represents changes in Earth's rotation rate; although this occurs, it is much too slow to be useful in seismology. The mode $_0T_2$ describes a twisting of the northern and southern hemispheres relative to each other; it has a period of about 44 minutes.

The first observations of free oscillations of the Earth were done during the great 1960 earthquake in Chile. Presently periods of thousands modes are known. These data are used for determining some large scale structures of the Earth interior.

P and S Waves in Earth's Mantle and Core

When an earthquake occurs, seismographs near the epicenter are able to record both P and S waves, but those at a greater distance no longer detect the high frequencies of the first S wave. Since shear waves cannot pass through liquids, this phenomenon was original evidence for the now well-established observation that the Earth has a liquid outer core, as demonstrated by Richard Dixon Oldham. This kind of observation has also been used to argue, by seismic testing, that the Moon has a solid core, although recent geodetic studies suggest the core is still molten.

Notation

The path that a wave takes between the focus and the observation point is often drawn as a ray diagram. An example of this is shown in a figure above. When reflections are taken into account there are an infinite number of paths that a wave can take. Each path is denoted by a set of letters that describe the trajectory and phase through the Earth. In general an upper case denotes a transmitted wave and a lower case denotes a reflected wave. The two exceptions to this seem to be "g" and "n".

c	the wave reflects off the outer core
d	a wave that has been reflected off a discontinuity at depth d
g	a wave that only travels through the crust
i	a wave that reflects off the inner core
I	a P-wave in the inner core
h	a reflection off a discontinuity in the inner core
J	an S wave in the inner core
K	a P-wave in the outer core
L	a Love wave sometimes called LT-Wave (Both caps, while an Lt is different)
n	a wave that travels along the boundary between the crust and mantle
P	a P wave in the mantle
p	a P wave ascending to the surface from the focus
R	a Rayleigh wave
S	an S wave in the mantle
s	an S wave ascending to the surface from the focus
w	the wave reflects off the bottom of the ocean
	No letter is used when the wave reflects off of the surfaces

For example:

- ScP is a wave that begins traveling towards the center of the Earth as an S wave. Upon reaching the outer core the wave reflects as a P wave.

- sPKIKP is a wave path that begins traveling towards the surface as an S-wave. At the surface it reflects as a P-wave. The P-wave then travels through the outer core, the inner core, the outer core, and the mantle.

Usefulness of P and S Waves in Locating an Event

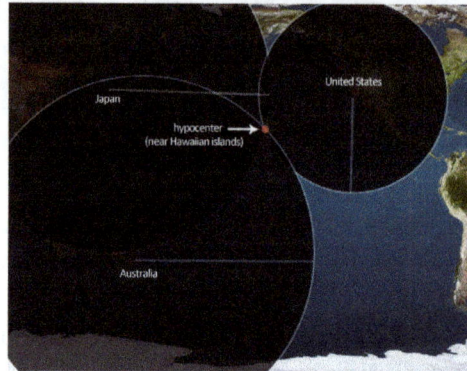

The Hypocenter/Epicenter of an earthquake is calculated by using the seismic data of that earthquake from at least three different locations.

In the case of local or nearby earthquakes, the difference in the arrival times of the P and S waves can be used to determine the distance to the event. In the case of earthquakes that have occurred at global distances, three or more geographically diverse observing stations (using a common clock) recording P-wave arrivals permits the computation of a unique time and location on the planet for the event. Typically, dozens or even hundreds of P-wave arrivals are used to calculate hypocenters. The misfit generated by a hypocenter calculation is known as "the residual". Residuals of 0.5 second or less are typical for distant events, residuals of 0.1-0.2 s typical for local events, meaning most reported P arrivals fit the computed hypocenter that well. Typically a location program will start by assuming the event occurred at a depth of about 33 km; then it minimizes the residual by adjusting depth. Most events occur at depths shallower than about 40 km, but some occur as deep as 700 km.

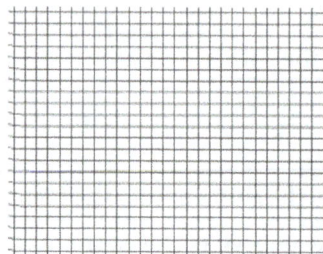

P- and S-waves sharing with the propagation

A quick way to determine the distance from a location to the origin of a seismic wave less than 200 km away is to take the difference in arrival time of the P wave and the S wave in seconds and multiply by 8 kilometers per second. Modern seismic arrays use more complicated earthquake location techniques.

At teleseismic distances, the first arriving P waves have necessarily travelled deep into the mantle, and perhaps have even refracted into the outer core of the planet, before travelling back up to the Earth's surface where the seismographic stations are located. The waves travel more quickly than if they had traveled in a straight line from the earthquake. This is due to the appreciably increased velocities within the planet, and is termed Huygens' Principle. Density in the planet increases with depth, which would slow the waves, but the modulus of the rock increases much more, so deeper means faster. Therefore, a longer route can take a shorter time.

The travel time must be calculated very accurately in order to compute a precise hypocenter. Since P waves move at many kilometers per second, being off on travel-time calculation by even a half second can mean an error of many kilometers in terms of distance. In practice, P arrivals from many stations are used and the errors cancel out, so the computed epicenter is likely to be quite accurate, on the order of 10–50 km or so around the world. Dense arrays of nearby sensors such as those that exist in California can provide accuracy of roughly a kilometer, and much greater accuracy is possible when timing is measured directly by cross-correlation of seismogram waveforms.

Types of Seismic Wave

P-wave

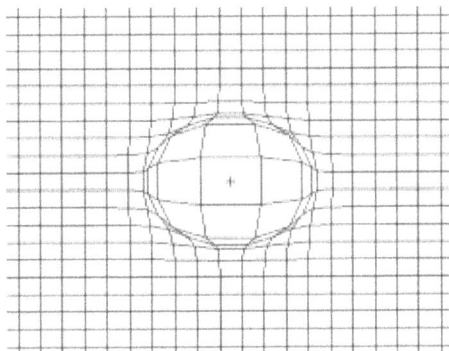

Representation of the propagation of a P-wave on a 2D grid (empirical shape)

P-waves are a type of elastic wave, and are one of the two main types of elastic body waves, called seismic waves in seismology, that travel through a continuum and are the first waves from an earthquake to arrive at a seismograph. The continuum is made up of gases (as sound waves), liquids, or solids, including the Earth. P-waves can be produced by earthquakes and recorded by seismographs. The name P-wave can stand for either pressure wave as it is formed from alternating compressions and rarefactions or primary wave, as it has the highest velocity and is therefore the first wave to be record-ed.

In isotropic and homogeneous solids, the mode of propagation of a P-wave is always longitudinal; thus, the particles in the solid vibrate along the axis of propagation (the direction of motion) of the wave energy.

Velocity

The velocity of P-waves in a homogeneous isotropic medium is given by

$$v_p = \sqrt{\frac{K + \frac{4}{3}\mu}{\rho}} = \sqrt{\frac{\lambda + 2\mu}{\rho}}$$

where K is the bulk modulus (the modulus of incompressibility), μ is the shear mod-ulus (modulus of rigidity, sometimes denoted as G and also called the second Lamé parameter), ρ is the density of the material through which the wave propagates, and λ is the first Lamé parameter.

Of these, density shows the least variation, so the velocity is mostly *controlled* by K and μ.

The elastic moduli P-wave modulus, M, is defined so that $M = K + 4\mu/3$ and thereby

$$v_p = \sqrt{M/\rho}$$

Typical values for P-wave velocity in earthquakes are in the range 5 to 8 km/s. The pre-cise speed varies according to the region of the Earth's interior, from less than 6 km/s in the Earth's crust to 13 km/s through the core.

Velocity of Common Rock Types		
Rocktype	**Velocity [m/s]**	**Velocity [ft/s]**
Unconsolidated Sandstone	4600 - 5200	15000 - 17000
Consolidated Sandstone	5800	19000
Shale	1800 - 4900	6000 -16000
Limestone	5800 - 6400	19000 - 21000
Dolomite	6400 - 7300	21000 - 24000
Anhydrite	6100	20000
Granite	5800 - 6100	19000 - 20000
Gabbro	7200	23600

Geologist Francis Birch discovered a relationship between the velocity of P waves and the density of the material the waves are traveling in:

$$V_p = a(\bar{M}) + b\rho$$

which later became known as Birch's law.

Seismic Waves in the Earth

Primary and secondary waves are body waves that travel within the Earth. The motion and behavior of both P-type and S-type in the Earth are monitored to probe the interior structure of the Earth. Discontinuities in velocity as a function of depth are indicative of changes in phase or composition. Differences in arrival times of waves originating in a seismic event like an earthquake as a result of waves taking different paths allow mapping of the Earth's inner structure.

P-wave Shadow Zone

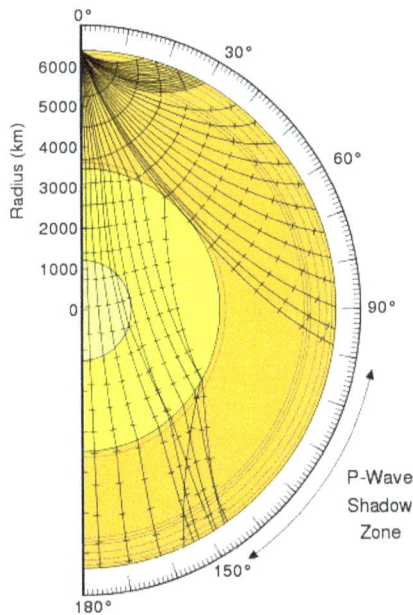

P-wave shadow zone (from USGS)

Almost all the information available on the structure of the Earth's deep interior is derived from observations of the travel times, reflections, refractions and phase transitions of seismic body waves, or normal modes. P-waves travel through the fluid layers of the Earth's interior, and yet they are refracted slightly when they pass through the transition between the semisolid mantle and the liquid outer core. As a result, there is a P-wave "shadow zone" between 103° and 142° from the earthquake's focus, where the initial P-waves are not registered on seismometers. In contrast, S-waves do not travel through liquids, rather, they are attenuated.

As an Earthquake Warning

Earthquake advance warning is possible by detecting the non-destructive primary waves that travel more quickly through the Earth's crust than do the destructive secondary and Rayleigh waves, in the same way that lightning flashes reach our eyes before we hear the thunder during a storm. The amount of advance warning depends on the delay between the arrival of the P-wave and other destructive waves, generally on the order of seconds up to about 60–90 seconds for deep, distant, large quakes such as Tokyo would have received before the 2011 Tohoku earthquake. The effectiveness of advance warning depends on accurate detection of the P-waves and rejection of ground vibrations caused by local activity (such as trucks or construction) as otherwise false-positive warnings will result. Earthquake Early Warning systems can be automated to allow for immediate safety actions such as issuing alerts, stopping elevators at the nearest floors or switching gas utilities off.

S-wave

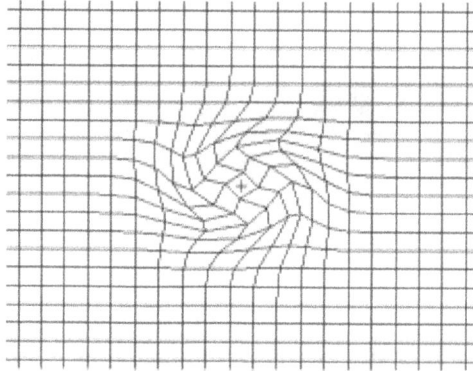

Propagation of a spherical S-wave in a 2d grid (empirical model)

In seismology, S-waves, secondary waves, or shear waves (sometimes called an elastic S-wave) are a type of elastic wave, and are one of the two main types of elastic body waves, so named because they move through the body of an object, unlike surface waves.

The S-wave moves as a shear or transverse wave, so motion is perpendicular to the direction of wave propagation. The wave moves through elastic media, and the main restoring force comes from shear effects. These waves do not diverge, and they obey the continuity equation for incompressible media:

$$\nabla \cdot \mathbf{u} = 0$$

Its name, S for secondary, comes from the fact that it is the second direct arrival on an earthquake seismogram, after the compressional primary wave, or P-wave, because S-waves travel slower in rock. Unlike the P-wave, the S-wave cannot travel through the molten outer core of the Earth, and this causes a shadow zone for S-waves opposite to where they originate. They can still appear in the solid inner core: when a P-wave strikes the boundary of molten and solid cores, S-waves will then propagate in the solid medium. And when the S-waves hit the boundary again they will in turn create P-waves. This property allows seismologists to determine the nature of the inner core.

As transverse waves, S-waves exhibit properties, such as polarization and birefringence, much like other transverse waves. S-waves polarized in the horizontal plane are classified as SH-waves. If polarized in the vertical plane, they are classified as SV-waves. When an S- or P-wave strikes an interface at an angle other than 90 degrees, a phenomenon known as mode conversion occurs. As described above, if the interface is between a solid and liquid, S becomes P or vice versa. However, even if the interface is between two solid media, mode conversion results. If a P-wave strikes an interface, four propagation modes may result: reflected and transmitted P and reflected and transmitted SV. Similarly, if an SV-wave strikes an interface, the same four modes occur in different proportions. The exact amplitudes of all these waves are described by the Zoeppritz equations, which in turn are solutions to the wave equation.

Theory

The prediction of S-waves came out of theory in the 1800s. Starting with the stress-strain relationship for an isotropic solid in Einstein notation:

$$\tau_{ij} = \lambda \delta_{ij} e_{kk} + 2\mu e_{ij}$$

where τ is the stress, λ and μ are the Lamé parameters (with μ as the shear modulus), δ_{ij} is the Kronecker delta, and the strain tensor is defined

$$e_{ij} = \frac{1}{2}\left(\partial_i u_j + \partial_j u_i\right)$$

for strain displacement u. Plugging the latter into the former yields:

$$\tau_{ij} = \lambda \delta_{ij} \partial_k u_k + \mu\left(\partial_i u_j + \partial_j u_i\right)$$

Newton's 2nd law in this situation gives the *homogeneous equation of motion* for seismic wave propagation:

$$\rho \frac{\partial^2 u_i}{\partial t^2} = \partial_j \tau_{ij}$$

where ρ is the mass density. Plugging in the above stress tensor gives:

$$\rho \frac{\partial^2 u_i}{\partial t^2} = \partial_i \lambda \partial_k u_k + \partial_j \mu\left(\partial_i u_j + \partial_j u_i\right)$$
$$= \lambda \partial_i \partial_k u_k + \mu \partial_i \partial_j u_j + \mu \partial_j \partial_j u_i$$

Applying vector identities and making certain approximations gives the seismic wave equation in homogeneous media:

$$\rho \ddot{u} = (\lambda + 2\mu)\nabla(\nabla \cdot u) - \mu\nabla \times (\nabla \times u)$$

where Newton's notation has been used for the time derivative. Taking the curl of this equation and applying vector identities eventually gives:

$$\nabla^2(\nabla \times \mathbf{u}) - \frac{1}{\beta^2}\frac{\partial^2}{\partial t^2}(\nabla \times \mathbf{u}) = 0$$

which is simply the wave equation applied to the curl of u with a velocity β satisfying

$$\beta^2 = \frac{\mu}{\rho}$$

This describes S-wave propagation. Taking the divergence of seismic wave equation in homogeneous media, instead of the curl, yields an equation describing P-wave propagation. The steady-state SH waves are defined by the Helmholtz equation

$$(\nabla^2 + k^2)\mathbf{u} = 0$$

where k is the wave number.

Rayleigh Wave

Rayleigh waves are a type of surface acoustic wave that travel on solids. They can be produced in materials in many ways, such as by a localized impact or by piezo-electric transduction, and are frequently used in non-destructive testing for detecting defects. They are part of the seismic waves that are produced on the Earth by earthquakes. When guided in layers they are referred to as Lamb waves, Rayleigh–Lamb waves, or generalized Rayleigh waves.

Characteristics

Rayleigh waves are a type of surface wave that travel near the surface of solids. Rayleigh waves include both longitudinal and transverse motions that decrease exponentially in amplitude as distance from the surface increases. There is a phase difference between these component motions.

Rayleigh Wave

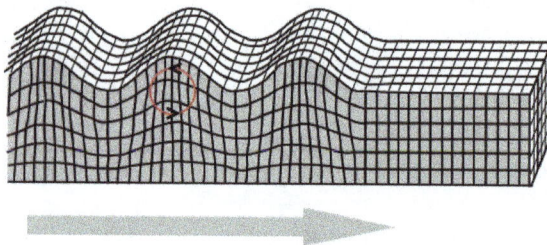

Picture of a Rayleigh wave.

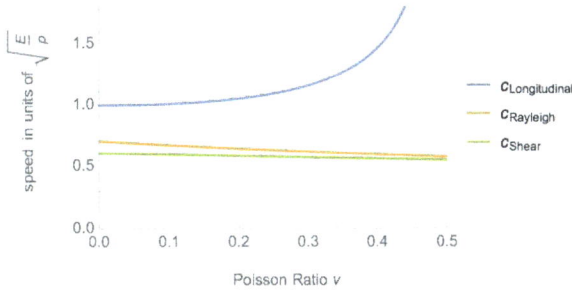

Comparison of the Rayleigh wave speed with shear and longitudinal wave speeds for an isotropic elastic material. The speeds are shown in dimensionless units.

The existence of Rayleigh waves was predicted in 1885 by Lord Rayleigh, after whom they were named. In isotropic solids these waves cause the surface particles to move in ellipses in planes normal to the surface and parallel to the direction of propagation – the major axis of the ellipse is vertical. At the surface and at shallow depths this motion is *retrograde*, that is the in-plane motion of a particle is counterclockwise when the wave travels from left to right. At greater depths the particle motion becomes *prograde*. In addition, the motion amplitude decays and the eccentricity changes as the depth into the material increases. The depth of significant displacement in the solid is approximately equal to the acoustic wavelength. Rayleigh waves are distinct from other types of surface or guided acoustic waves such as Love waves or Lamb waves, both being types of guided waves supported by a layer, or longitudinal and shear waves, that travel in the bulk.

Rayleigh waves have a speed slightly less than shear waves by a factor dependent on the elastic constants of the material. The typical speed of Rayleigh waves in metals is of the order of 2–5 km/s, the typical Rayleigh speed in the ground is of the order of 50–300 m/s. For linear elastic materials with positive Poisson ratio ($\nu > 0$), the Rayleigh wave speed can be approximated as $c_R / c_S = \dfrac{0.862 + 1.14\nu}{1+\nu}$.. Since Rayleigh waves are confined near the surface, their in-plane amplitude when generated by a point source decays only as $1/\sqrt{r}$, where r is the radial distance. Surface waves therefore decay more slowly with distance than do bulk waves, which spread out in three dimensions from a point source.

In seismology, Rayleigh waves (called "ground roll") are the most important type of surface wave, and can be produced (apart from earthquakes), for example, by ocean waves, by explosions, by railway trains and ground vehicles, or by a sledgehammer impact.

Rayleigh Wave Dispersion

The elastic constants often change with depth, due to the changing properties of the material. This means that the velocity of a Rayleigh wave becomes dependent on the wavelength (and therefore frequency), a phenomenon referred to as dispersion. Waves affected by dispersion have a different wave train shape. Rayleigh waves on ideal, homogeneous and flat elastic solids show no dispersion. However, if a solid or structure has a density or sound velocity that varies with depth, Rayleigh waves become dispersive. One example is Rayleigh waves on the Earth's surface: those waves with a higher frequency travel more slowly than those with a lower frequency. This occurs because a Rayleigh wave of lower frequency has a relatively long wavelength. The displacement of long wavelength waves penetrates more deeply into the Earth than short wavelength waves. Since the speed of waves in the Earth increases with increasing depth, the longer wavelength (low frequency) waves can travel faster than the shorter wavelength (high frequency) waves. Rayleigh waves thus often appear spread out on seismograms recorded at distant earthquake recording stations. It is also possible to observe Rayleigh wave dispersion in thin films or multi-layered structures.

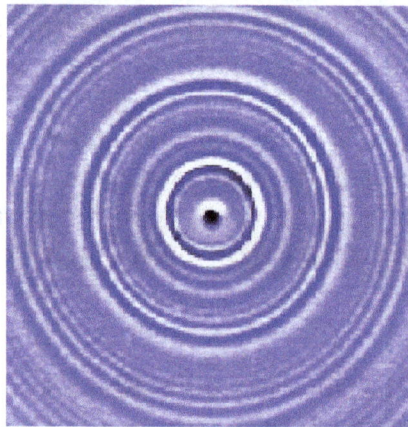

Dispersion of Rayleigh waves in a thin gold film on glass.

Rayleigh Waves in Non-destructive Testing

Rayleigh waves are widely used for materials characterization, to discover the mechanical and structural properties of the object being tested – like the presence of cracking, and the related shear modulus. This is in common with other types of surface waves. The Rayleigh waves used for this purpose are in the ultrasonic frequency range.

They are used at different length scales because they are easily generated and detected on the free surface of solid objects. Since they are confined in the vicinity of the free surface within a depth (~ the wavelength) linked to the frequency of the wave, different frequencies can be used for characterization at different length scales.

Rayleigh Waves in Electronic Devices

Rayleigh waves propagating at high ultrasonic frequencies (10-1000 MHz) are used widely in different electronic devices. In addition to Rayleigh waves, some other types of surface acoustic waves (SAW), e.g. Love waves, are also used for this purpose. Examples of electronic devices using Rayleigh waves are filters, resonators, oscillators, sensors of pressure, temperature, humidity, etc. Operation of SAW devices is based on the transformation of the initial electric signal into a surface wave that, after achieving the required changes to the spectrum of the initial electric signal as a result of its interaction with different types of surface inhomogeneity, is transformed back into a modified electric signal. The transformation of the initial electric energy into mechanical energy (in the form of SAW) and back is usually accomplished via the use of piezoelectric materials for both generation and reception of Rayleigh waves as well as for their propagation.

Rayleigh Waves in Geophysics

Rayleigh Waves from Earthquakes

Because Rayleigh waves are surface waves, the amplitude of such waves generated by an earthquake generally decreases exponentially with the depth of the hypocenter (focus). However, large earthquakes may generate Rayleigh waves that travel around the Earth several times before dissipating.

In seismology longitudinal and shear waves are known as P-waves and S-waves, respectively, and are termed body waves. Rayleigh waves are generated by the interaction of P- and S- waves at the surface of the earth, and travel with a velocity that is lower than the P-, S-, and Love wave velocities. Rayleigh waves emanating outward from the epicenter of an earthquake travel along the surface of the earth at about 10 times the speed of sound in air (0.340 km/s), that is ~3 km/s.

Due to their higher speed, the P- and S-waves generated by an earthquake arrive before the surface waves. However, the particle motion of surface waves is larger than that of body waves, so the surface waves tend to cause more damage. In the case of Rayleigh waves, the motion is of a rolling nature, similar to an ocean surface wave. The intensity of Rayleigh wave shaking at a particular location is dependent on several factors:

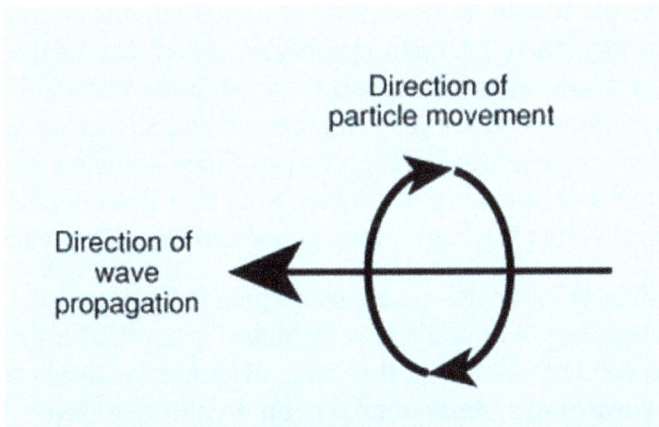

Rayleigh wave direction

- The size of the earthquake.

- The distance to the earthquake.

- The depth of the earthquake.

- The geologic structure of the crust.

- The focal mechanism of the earthquake.

- The rupture directivity of the earthquake.

Local geologic structure can serve to focus or defocus Rayleigh waves, leading to significant differences in shaking over short distances.

Rayleigh Waves in Seismology

Low frequency Rayleigh waves generated during earthquakes are used in seismology to characterise the Earth's interior. In intermediate ranges, Rayleigh waves are used in geophysics and geotechnical engineering for the characterisation of oil deposits. These applications are based on the geometric dispersion of Rayleigh waves and on the solution of an inverse problem on the basis of seismic data collected on the ground surface using active sources (falling weights, hammers or small explosions, for example) or by recording microtremors. Rayleigh ground waves are important also for environmental noise and vibration control since they make a major contribution to traffic-induced ground vibrations and the associated structure-borne noise in buildings.

Other Manifestations

Animals

Low frequency (< 20 Hz) Rayleigh waves are inaudible, yet they can be detected by many mammals, birds, insects and spiders. Humans should be able to detect such Rayleigh waves through their Pacinian corpuscles, which are in the joints, although people do not seem to consciously respond to the signals. Some animals seem to use Rayleigh waves to communicate. In particular, some biologists theorize that elephants may use vocalizations to generate Rayleigh waves. Since Rayleigh waves decay slowly, they should be detectable over long distances. Note that these Rayleigh waves have a much higher frequency than Rayleigh waves generated by earthquakes.

After the 2004 Indian Ocean earthquake, some people have speculated that Rayleigh waves served as a warning to animals to seek higher ground, allowing them to escape the more slowly traveling tsunami. At this time, evidence for this is mostly anecdotal. Other animal early warning systems may rely on an ability to sense infrasonic waves traveling through the air.

Love Wave

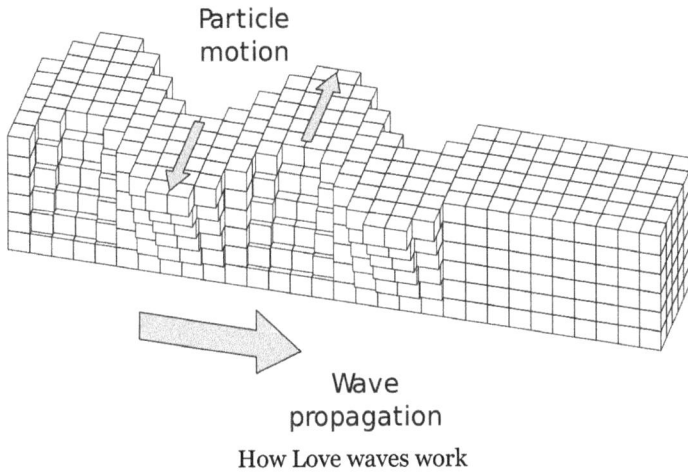

How Love waves work

In elastodynamics, Love waves, named after Augustus Edward Hough Love, are horizontally polarized surface waves. The Love wave is a result of the interference of many shear waves (S–waves) guided by an elastic layer, which is welded to an elastic half space on one side while bordering a vacuum on the other side. In seismology, Love waves (also known as Q waves (Quer: German for lateral)) are surface seismic waves that cause horizontal shifting of the Earth during an earthquake. Augustus Edward Hough Love predicted the existence of Love waves mathematically in 1911. They form a distinct class, different from other types of seismic waves, such as P-waves and S-waves (both body waves), or Rayleigh waves (another type of surface wave). Love waves travel with a lower velocity than P- or S- waves, but faster than Rayleigh waves. These waves are observed only when there is a low velocity layer overlying a high velocity layer/sub–layers.

Description

The particle motion of a Love wave forms a horizontal line perpendicular to the direction of propagation (i.e. are transverse waves). Moving deeper into the material, motion can decrease to a "node" and then alternately increase and decrease as one examines deeper layers of particles. The amplitude, or maximum particle motion, often decreases rapidly with depth.

Since Love waves travel on the Earth's surface, the strength (or amplitude) of the waves decrease exponentially with the depth of an earthquake. However, given their confine ment to the surface, their amplitude decays only as $\frac{1}{\sqrt{r}}$, where r represents the distance the wave has travelled from the earthquake. Surface waves therefore decay more slowly with distance than do body waves, which travel in three dimensions. Large earthquakes may generate Love waves that travel around the Earth several times before dissipating.

Since they decay so slowly, Love waves are the most destructive outside the immediate area of the focus or epicentre of an earthquake. They are what most people feel directly during an earthquake.

In the past, it was often thought that animals like cats and dogs could predict an earthquake before it happened. However, they are simply more sensitive to ground vibrations than humans and able to detect the subtler body waves that precede Love waves, like the P-waves and the S-waves.

Basic Theory

The conservation of linear momentum of a linear elastic material can be written as

$$\nabla \cdot (C : \nabla \mathbf{u}) = \rho \ddot{\mathbf{u}}$$

where $\mathbf{u}$ is the displacement vector and C is the stiffness tensor. Love waves are a special solution ($\mathbf{u}$) that satisfy this system of equations. We typically use a Cartesian coordinate system (x, y, z) to describe Love waves.

Consider an isotropic linear elastic medium in which the elastic properties are functions of only the z coordinate, i.e., the Lamé parameters and the mass density can be expressed as $\lambda(z), \mu(z), \rho(z)$. Displacements (u, v, w) produced by Love waves as a function of time (t) have the form

$$u(x, y, z, t) = 0, \quad v(x, y, z, t) = \hat{v}(x, z, t), \quad w(x, y, z, t) = 0.$$

These are therefore antiplane shear waves perpendicular to the (x, z) plane. The function $\hat{v}(x, z, t)$ can be expressed as the superposition of harmonic waves with varying wave numbers (k) and frequencies (ω). Let us consider a single harmonic wave, i.e.,

$$\hat{v}(x, z, t) = V(k, z, \omega) \exp[i(kx - \omega t)]$$

where $i = \sqrt{-1}$. The stresses caused by these displacements are

$$\sigma_{xx} = 0, \quad \sigma_{yy} = 0, \quad \sigma_{zz} = 0, \quad \tau_{zx} = 0,$$

$$\tau_{yz} = \mu(z) \frac{dV}{dz} \exp[i(kx - \omega t)],$$

$$\tau_{xy} = ik\mu(z)V(k, z, \omega) \exp[i(kx - \omega t)].$$

If we substitute the assumed displacements into the equations for the conservation of momentum, we get a simplified equation

$$\frac{d}{dz}\left[\mu(z)\frac{dV}{dz}\right] = [k^2 \mu(z) - \omega^2 \rho(z)]V(k, z, \omega).$$

The boundary conditions for a Love wave are that the surface tractions at the free surface $(z=0)$ must be zero. Another requirement is that the stress component τ_{yz} in a layer medium must be continuous at the interfaces of the layers. To convert the second order differential equation in V into two first order equations, we express this stress component in the form

$$\tau_{yz} = T(k,z,\omega)\exp[i(kx-\omega t)]$$

to get the first order conservation of momentum equations

$$\frac{d}{dz}\begin{bmatrix} V \\ T \end{bmatrix} = \begin{bmatrix} 0 & 1/\mu(z) \\ k^2\mu(z)-\omega^2\rho(z) & 0 \end{bmatrix}\begin{bmatrix} V \\ T \end{bmatrix}.$$

The above equations describe an eigenvalue problem whose solution eigenfunctions can be found by a number of numerical methods. Another common, and powerful, approach is the propagator matrix method (also called the matricant approach)

Ductility (Earth Science)

Fig. - A vertical viewpoint of a rock outcrop that has undergone ductile deformation to create a series of asymmetric folds.

In Earth science, as opposed to Materials Science, Ductility refers to the capacity of a rock to deform to large strains without macroscopic fracturing. Such behavior may occur in unlithified or poorly lithified sediments, in weak materials such as halite or at greater depths in all rock types where higher temperatures promote crystal plasticity and higher confining pressures suppress brittle fracture. In addition, when a material is behaving ductilely, it exhibits a linear stress vs strain relationship past the elastic limit.

Ductile deformation is typically characterized by diffuse deformation (i.e. lacking a discrete fault plane) and on a stress-strain plot is accompanied by steady state sliding at failure, compared to the sharp stress drop observed in experiments during brittle failure.

Brittle-Ductile Transition Zone

The brittle-ductile transition zone is characterized by a change in rock failure mode, at an approximate average depth of 10-15 km (~ 6.2-9.3 miles) in continental crust, below which rock becomes less likely to fracture and more likely to deform ductilely. The zone exists because as depth increases confining pressure increases, and brittle strength increases with confining pressure whilst ductile strength decreases with increasing temperature. The transition zone occurs at the point where brittle strength equals ductile strength. In glacial ice this zone is at approximately 30 m (100 ft) depth.

Not all materials, however, abide by this transition. It is possible and not rare for material above the transition zone to deform ductilely, and for material below to deform in a brittle manner. The depth of the material does exert an influence on the mode of deformation, but other substances, such as loose soils in the upper crust, malleable rocks, biological debris, and more are just a few examples of that which does not deform in accordance to the transition zone.

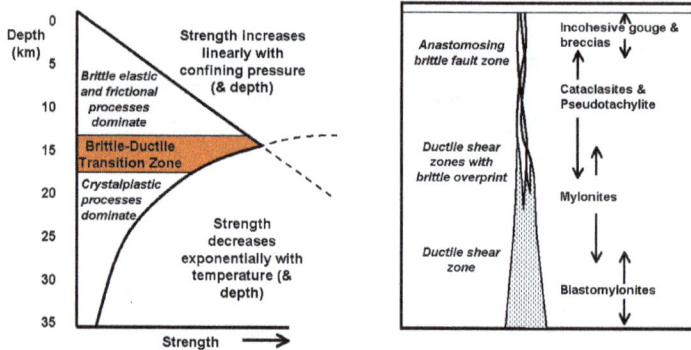

Fig. - A generalized diagram of the deformation mechanisms and structural formations that dominate at certain depths within the Earth's crust.

The type of dominating deformation process also has a great impact on the types of rocks and structures found at certain depths within the Earth's crust. As evident from Fig., different geological formations and rocks are found in accordance to the dominant deformation process. Gouge and Breccia form in the uppermost, brittle regime while Cataclasite and Pseudotachylite form in the lower parts of the brittle regime, edging upon the transition zone. Mylonite forms in the more ductile regime at greater depths while Blastomylonite forms well past the transition zone and well into the ductile regime, even deeper into the crust.

Quantification

Ductility is a material property that can be expressed in a variety of ways. Mathematically, it is commonly expressed as a *total quantity of elongation* or a *total quantity of the change in cross sectional area* of a specific rock until macroscopic brittle behavior, such as fracturing, is observed. For accurate measurement, this must be done under several controlled conditions, including but not limited to Pressure, Temperature, Moisture Content, Sample Size, etc., for all can impact the measured ductility. It is important to understand that even the same type of rock or mineral may exhibit different behavior and degrees of ductility due to internal heterogeneities small scale differences between each individual sample. The two quantities are expressed in the form of a ratio or a percent.

% Elongation of a Rock = $\%\Delta l = \dfrac{l_f - l_i}{l_i} \times 100$

Where:

l_i = Initial Length of Rock

l_f = Final Length of Rock

% Change in Area of a Rock = $\%\Delta A = \dfrac{A_f - A_i}{A_i} \times 100$

Where:

A_i = Initial Area

A_f = Final Area

For each of these methods of quantifying, one must take measurements of both the initial and final dimensions of the rock sample. For Elongation, the measurement is a uni-dimensional initial and final length, the former measured before any Stress is applied and the latter measuring the length of the sample after fracture occurs. For Area, it is strongly preferable to use a rock that has been cut into a cylindrical shape before stress application so that the cross-sectional area of the sample can be taken.

Cross-Sectional Area of a Cylinder = Area of a Circle = $A = \pi r^2$

Using this, the initial and final areas of the sample can be used to quantify the % change in the area of the rock.

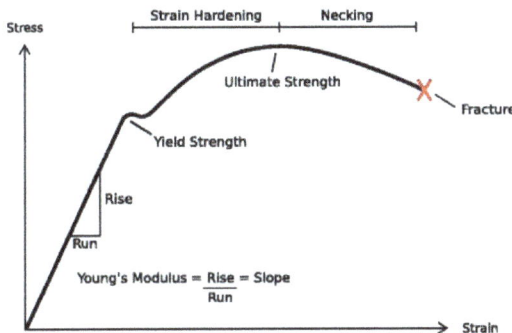

Fig. - Stress vs Strain Curve displaying both ductile and brittle deformation behavior.

Deformation

Any material is shown to be able to deform ductilely or brittlely, in which the type of deformation is governed by both the external conditions around the rock and the internal conditions sample. External conditions include temperature, confining pressure, presence of fluids, etc. while internal conditions include the arrangement of the crystal lattice, the chemical composition of the rock sample, the grain size of the material, etc.

Ductilely Deformative behavior can be grouped into three categories: Elastic, Viscous, and Crystal-Plastic Deformation.

Elastic Deformation

Elastic Deformation is deformation which exhibits a linear stress-strain relationship (quantified by Young's Modulus) and is derived from Hooke's Law of spring forces. In elastic deformation, objects show no permanent deformation after the stress has been removed from the system and return to their original state.

$$\sigma = E\dot{o}$$

Where:

σ = Stress (In Pascals)

E = Young's Modulus (In Pascals)

ϵ = Strain (Unitless)

Viscous Deformation

Viscous Deformation is when rocks behave and deform more like a fluid than a solid. This often occurs under great amounts of pressure and at very high temperatures. In viscous deformation, stress is proportional to the strain rate, and each rock sample has its own material property called its Viscosity. Unlike elastic deformation, viscous deformation is permanent even after the stress has been removed.

$$\sigma = \eta\xi$$

Where:

σ = Stress (In Pascals)

η = Viscosity (In Pascals * Seconds)

ξ = Strain Rate (In 1/Seconds)

Crystal-Plastic Deformation

Crystal-Plastic Deformation occurs at the atomic scale and is governed by its own set of specific mechanisms that deform crystals by the movements of atoms and atomic planes through the crystal lattice. Like viscous deformation, it is also a permanent form of deformation. Mechanisms of crystal-plastic deformation include Pressure solution, Dislocation creep, and Diffusion creep.

Biological Materials

In addition to rocks, biological materials such as wood, lumber, bone, etc. can be assessed for their ductility as well, for many behave in the same manner and possess the same characteristics as abiotic Earth materials. This assessment was done in Hiroshi Yoshihara's experiment, "Plasticity Analysis of the Strain in the Tangential Direction of Solid Woo Subjected to Compression Load in the Longitudinal Direction." The study aimed to analyze the behavioral rheology of 2 wood specimens, the Sitka Spruce and Japanese Birch. In the past, it was shown that solid wood, when subjected to compressional stresses, initially has a linear stress-strain diagram (indicative of elastic deformation) and later, under greater load, demonstrates a non-linear diagram indicative of ductile objects. To analyze the rheology, the stress was restricted to uniaxial compression in the longitudinal direction and the post-linear behavior was analyzed using plasticity theory. Controls included moisture content in the lumber, lack of defects such as knots or grain distortions, temperature at 20 C, relative humidity at 65%, and size of the cut shapes of the wood samples.

Results obtained from the experiment exhibited a linear stress-strain relationship during elastic deformation but also an unexpected non-linear relationship between stress and strain for the lumber after the elastic limit was reached, deviating from the model of plasticity theory. Multiple reasons were suggested as to why this came about. First, since wood is a biological material, it was suggested that under great stress in the experiment, the crushing of cells within the sample could have been a cause for deviation from perfectly plastic behavior. With greater destruction of cellular material, the stress-strain relationship is hypothesized to become more and more nonlinear and non-ideal with greater stress. Additionally, because the samples were inhomogeneous (non-uniform) materials, it was assumed that some bending or distortion may have occurred in the samples that could have deviated the stress from being perfectly uniaxial. This may have also been induced by other factors like irregularities in the cellular density profile and distorted sample cutting.

The conclusions of the research accurately showed that although biological materials can behave like rocks undergoing deformation, there are many other factors and variables that must be considered, making it difficult to standardize the ductility and material properties of a biological substance.

Peak Ductility Demand

Peak Ductility Demand is a quantity used particularly in the fields of architecture, geological engineering, and mechanical engineering. It is defined as the amount of ductile deformation a material must be able to withstand (when exposed to a stress) without brittle fracture or failure. This quantity is particularly useful in the analysis of failure of structures in response to earthquakes and seismic waves.

It has been shown that earthquake aftershocks can increase the peak ductility demand with respect to the mainshocks by up to 10%.

Seismic Tomography

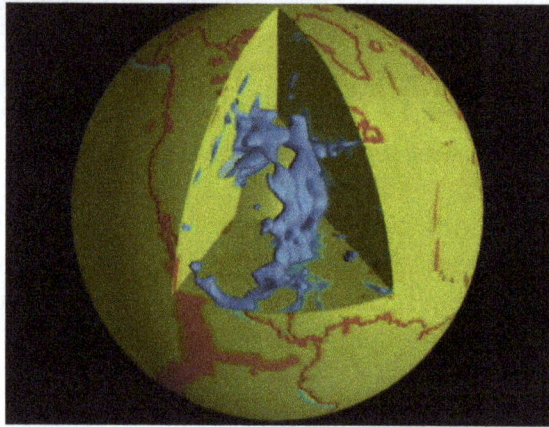

NASA tomographic image of the subducted Farallon Plate in the mantle beneath eastern North America.

Seismic tomography is a technique for imaging the subsurface of the Earth with seismic waves produced by earthquakes or explosions. P-, S-, and surface waves can be used for tomographic models of different resolutions based on seismic wavelength, wave source distance, and the seismograph array coverage. The data received at seismometers are used to solve an inverse problem, wherein the locations of reflection and refraction of the wave paths are determined. This solution can be used to create 3D images of velocity anomalies which may be interpreted as structural, thermal, or compositional variations. Geoscientists use these images to better understand core, mantle, and plate tectonic processes.

Theory

Tomography is solved as an inverse problem. Seismic travel time data are compared to an initial Earth model and the model is modified until the best possible fit between the model predictions and observed data is found. Seismic waves would travel in straight lines if Earth was of uniform composition, but the compositional layering, tectonic structure, and thermal variations reflect and refract seismic waves. The location and magnitude of these variations can be calculated by the inversion process, although solutions to tomographic inversions are non-unique.

Seismic tomography is similar to medical x-ray computed tomography (CT scan) in that a computer processes receiver data to produce a 3D image, although CT scans use attenuation instead of traveltime difference. Seismic tomography has to deal with the analysis of curved ray paths which are reflected and refracted within the earth and potential uncertainty in the location of the earthquake hypocenter. CT scans use linear x-rays and a known source.

History

Seismic tomography requires large datasets of seismograms and well-located earthquake or explosion sources. These became more widely available in the 1960s with the expansion of global seismic networks and in the 1970s when digital seismograph data archives were established. These developments occurred concurrently with advancements in computing power that were required to solve inverse problems and generate theoretical seismograms for model testing.

In 1977, P-wave delay times were used to create the first seismic array-scale 2D map of seismic velocity. In the same year, P-wave data were used to determine 150 spherical harmonic coefficients for velocity anomalies in the mantle. The first model using iterative techniques, required when there are a large numbers of unknowns, was done in 1984. This built upon the first radially anisotropic model of the Earth, which provided the required initial reference frame to compare tomographic models to for iteration. Initial models had resolution of ~3000 to 5000 km, as compared to the few hundred kilometer resolution of current models.

Seismic tomographic models improve with advancements in computing and expansion of seismic networks. Recent models of global body waves used over 10^7 traveltimes to model 10^5 to 10^6 unknowns.

Process

Seismic tomography uses seismic records to create 2D and 3D images of subsurface anomalies by solving large inverse problems such that generate models consistent with observed data. Various methods are used to resolve anomalies in the crust and lithosphere, shallow mantle, whole mantle, and core based on the availability of data and types of seismic waves that penetrate the region at a suitable wavelength for feature resolution. The accuracy of the model is limited by availability and accuracy of seismic data, wave type utilized, and assumptions made in the model.

P-wave data are used in most local models and global models in areas with sufficient earthquake and seismograph density. S- and surface wave data are used in global models when this coverage is not sufficient, such as in ocean basins and away from subduction zones. First-arrival times are the most widely used, but models utilizing reflected and refracted phases are used in more complex models, such as those imaging the core. Differential traveltimes between wave phases or types are also used.

Local Tomography

Local tomographic models are often based on a temporary seismic array targeting specific areas, unless in a seismically active region with extensive permanent network coverage. These allow for the imaging of the crust and upper mantle.

- *Diffraction and wave equation tomography* use the full waveform, rather than just the first arrival times. The inversion of amplitude and phases of all arrivals provide more detailed density information than transmission traveltime alone. Despite the theoretical appeal, these methods are not widely employed because of the computing expense and difficult inversions.

- *Reflection tomography* originated with exploration geophysics. It uses an artificial source to resolve small-scale features at crustal depths. *Wide-angle tomography* is similar, but with a wide source to receiver offset. This allows for the detection of seismic waves refracted from sub-crustal depths and can determine continental architecture and details of plate margins. These two methods are often used together.

- *Local earthquake tomography* is used in seismically active regions with sufficient seismometer coverage. Given the proximity between source and receivers, a precise earthquake focus location must be known. This requires the simultaneous iteration of both structure and focus locations in model calculations.

- *Teleseismic tomography* uses waves from distant earthquakes that deflect upwards to a local seismic array. The models can reach depths similar to the array aperture, typically to depths for imaging the crust and lithosphere (a few hundred kilometers). The waves travel near 30° from vertical, creating a vertical distortion to compact features.

Regional or Global Tomography

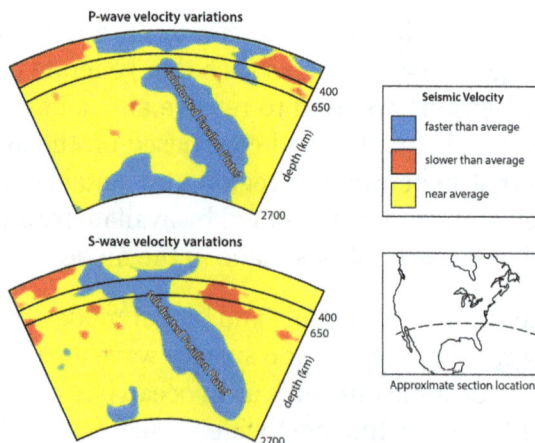

Simplified and interpreted P- and S-wave velocity variations in the mantle across southern North America showing the subducted Farallon Plate.

Regional to global scale tomographic models are generally based on long wavelengths. Various models have better agreement with each other than local models due to the large feature size they image, such as subducted slabs and superplumes. The trade off from whole mantle to whole earth coverage is the coarse resolution (hundreds of kilometers) and difficulty imaging small features (e.g. narrow plumes). Although often used to image different parts of the subsurface, P- and S-wave derived models broadly agree where there is image overlap. These models use data from both permanent seismic stations and supplementary temporary arrays.

- First arrival traveltime *P-wave* data are used to generate the highest resolution tomographic images of the mantle. These models are limited to regions with sufficient seismograph coverage and earthquake density, therefore cannot be used for areas such as inactive plate interiors and ocean basins without seismic networks. Other phases of P-waves are used to image the deeper mantle and core.

- In areas with limited seismograph or earthquake coverage, multiple phases of *S-waves* can be used for tomographic models. These are of lower resolution than P-wave models, due to the distances involved and fewer bounce-phase data available. S-waves can also be used in conjunction with P-waves for differential arrival time models.

- *Surface waves* can be used for tomography of the crust and upper mantle where no body wave (P and S) data are available. Both Rayleigh and Love waves can be used. The low frequency waves lead to low resolution models, therefore these models have difficulty with crustal structure. *Free oscillations*, or normal mode seismology, are the long wavelength, low frequency movements of the surface of the earth which can be thought of as a type of surface wave. The frequencies of these oscillations can be obtained through Fourier transformation of seismic data. The models based on this method are of broad scale, but have the advantage of relatively uniform data coverage as compared to data sourced directly from earthquakes.

- *Attenuation tomography* attempts to extract the anaelastic signal from the elastic-dominated waveform of seismic waves. The advantage of this method is its sensitivity to temperature, thus ability to image thermal features such as mantle plumes and subduction zones. Both surface and body waves have been used in this approach.

- *Ambient noise tomography* cross-correlates waveforms from random wavefields generated by oceanic and atmospheric disturbances which subsequently become diffuse within the Earth. This method has produced high-resolution images and is an area of active research.

- Waveforms are modeled as rays in seismic analysis, but all waves are affected by the material near the ray path. The finite frequency effect is the result the surrounding medium has on a seismic record. *Finite frequency tomography* accounts for this in determining both travel time and amplitude anomalies, increasing image resolution. This has the ability to resolve much larger variations (i.e. 10-30%) in material properties.

Applications

Seismic tomography can resolve anisotropy, anelasticity, density, and bulk sound density. Variations in these parameters may be a result of thermal or chemical differences, which are attributed to processes such as mantle plumes, subducting slabs, and mineral phase changes. Larger scale features that can be imaged with tomography include the high velocities beneath continental shields and low velocities under ocean spreading centers.

Hotspots

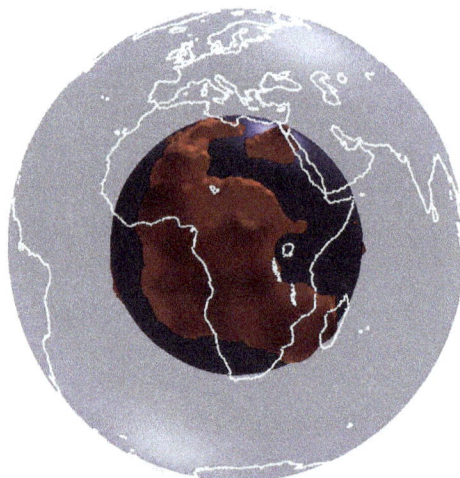

Tomographic image of the African large low-shear-velocity province (superplume).

The mantle plume hypothesis proposes that areas of volcanism not readily explained by plate tectonics, called hotspots, are a result of thermal upwelling from as deep as the core-mantle boundary that become diapirs in the crust. This is an actively contested theory, although tomographic images suggest there are anomalies beneath some hotspots. The best imaged of these are large low-shear-velocity provinces, or superplumes, visible on S-wave models of the lower mantle and believed to reflect both thermal and compositional differences.

The Yellowstone hotspot is responsible for volcanism at the Yellowstone Caldera and a series of extinct calderas along the Snake River Plain. The Yellowstone Geodynamic Project sought to image the plume beneath the hotspot. They found a strong low-velocity body from ~30 to 250 km depth beneath Yellowstone and a weaker anomaly from 250 to 650 km depth which dipped 60° west-northwest. The authors attribute these features to the mantle plume beneath the hotspot being deflected eastward by flow in the upper mantle seen in S-wave models.

The Hawaii hotspot lies beneath the Hawaiian Islands and Emperor Seamounts. Tomographic images show it to be 500 to 600 km wide and up to 2,000 km deep.

Subduction Zones

Subducting plates are colder than the mantle into which they are moving. This creates a fast anomaly that is visible in tomographic images. Both the Farallon plate that subducted beneath the west coast of North America and the northern portion of the Indian plate that has subducted beneath Asia have been imaged with tomography.

Limitations

Global seismic networks have expanded steadily since the 1960s, but are still concentrated on continents and in seismically active regions. Oceans, particularly in the southern hemisphere, are under-covered. Tomographic models in these areas will improve when more data becomes available. The uneven distribution of earthquakes naturally biases models to better resolution in seismically active regions.

The type of wave used in a model limits the resolution it can achieve. Longer wavelengths are able to penetrate deeper into the earth, but can only be used to resolve large features. Finer resolution can be achieved with surface waves, with the trade off that they cannot be used in models of the deep mantle. The disparity between wavelength and feature scale causes anomalies to appear of reduced magnitude and size in images. P- and S-wave models respond differently to the types of anomalies depending on the driving material property. First arrival time based models naturally prefer faster pathways, causing models based on these data to have lower resolution of slow (often hot) features. Shallow models must also consider the significant lateral velocity variations in continental crust.

Seismic tomography provides only the current velocity anomalies. Any prior structures are unknown and the slow rates of movement in the subsurface (mm to cm per year) prohibit resolution of changes over modern timescales.

Tomographic solutions are non-unique. Although statistical methods can be used to analyze the validity of a model, unresolvable uncertainty remains. This contributes to difficulty comparing the validity of different model results.

Computing power limits the amount of seismic data, number of unknowns, mesh size, and iterations in tomographic models. This is of particular importance in ocean basins, which due to limited network coverage and earthquake density require more complex processing of distant data. Shallow oceanic models also require smaller model mesh size due to the thinner crust.

Tomographic images are typically presented with a color ramp representing the strength of the anomalies. This has the consequence of making equal changes appear of differing magnitude based on visual perceptions of color, such as the change from orange to red being more subtle than blue to yellow. The degree of color saturation can also visually skew interpretations. These factors should be considered when analyzing images.

Hotspot (Geology)

Diagram showing a cross section though the Earth's lithosphere (in yellow) with magma rising from the mantle (in red)

In geology, the places known as hotspots or hot spots are volcanic regions thought to be fed by underlying mantle that is anomalously hot compared with the surrounding mantle. They may be on, near to, or far from tectonic plate boundaries. Currently, there are two hypotheses that attempt to explain their origins. One suggests that hotspots are due to mantle plumes that rise as thermal diapirs from the core–mantle boundary. The other hypothesis is that lithospheric extension permits the passive rising of melt from shallow depths. This hypothesis considers the term "hotspot" to be a misnomer, asserting that the mantle source beneath them is, in fact, not anomalously hot at all. Well known examples include Hawaii and Yellowstone.

Background

The origins of the concept of hotspots lie in the work of J. Tuzo Wilson, who postulated in 1963 that the Hawaiian Islands result from the slow movement of a tectonic plate across a hot region beneath the surface. It was later postulated that hotspots are fed by narrow streams of hot mantle rising from the Earth's core–mantle boundary in a structure called a mantle plume. Whether or not such mantle plumes exist is currently the subject of a major controversy in Earth science. Estimates for the number of hotspots postulated to be fed by mantle plumes has ranged from about 20 to several thousands, over the years, with most geologists considering a few tens to exist. Hawaii, Réunion, Yellowstone, Galápagos, and Iceland are some of the currently most active volcanic regions to which the hypothesis is applied.

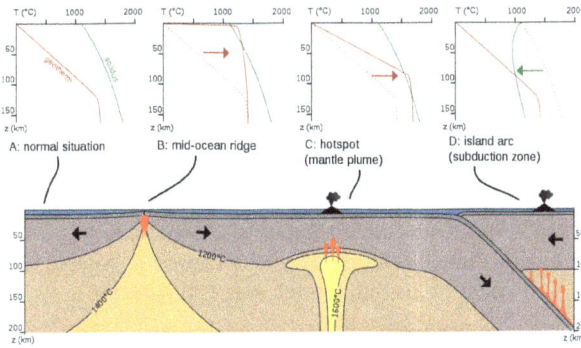

Schematic diagram showing the physical processes inside the Earth that lead to the generation of magma. Partial melting begins above the fusion point.

Most hotspot volcanoes are basaltic (e.g., Hawaii, Tahiti). As a result, they are less explosive than subduction zone volcanoes, in which water is trapped under the overriding plate. Where hotspots occur in continental regions, basaltic magma rises through the continental crust, which melts to form rhyolites. These rhyolites can form violent eruptions. For example, the Yellowstone Caldera was formed by some of the most powerful volcanic explosions in geologic history. However, when the rhyolite is completely erupted, it may be followed by eruptions of basaltic magma rising through the same lithospheric fissures (cracks in the lithosphere). An example of this activity is the Ilgachuz Range in British Columbia, which was created by an early complex series of trachyte and rhyolite eruptions, and late extrusion of a sequence of basaltic lava flows.

The hotspot hypothesis is now closely linked to the mantle plume hypothesis.

Comparison with Island Arc Volcanoes

Hotspot volcanoes are considered to have a fundamentally different origin from island arc volcanoes. The latter form over subduction zones, at converging plate boundaries. When one oceanic plate meets another, the denser plate is forced downward into a deep ocean trench. This plate, as it is subducted, releases water into the base of the over-riding plate, and this water mixes with the rock, thus changing its composition causing some rock to melt and rise. It is this that fuels a chain of volcanoes, such as the Aleutian Islands, near Alaska.

Hotspot Volcanic Chains

The joint mantle plume/hotspot hypothesis envisages the feeder structures to be fixed relative to one another, with the continents and seafloor drifting overhead. The hypothesis thus predicts that time-progressive chains of volcanoes are developed on the surface. Examples are Yellowstone, which lies at the end of a chain of extinct calderas, which become progressively older to the west. Another example is the Hawaiian archipelago, where islands become progressively older and more deeply eroded to the northwest.

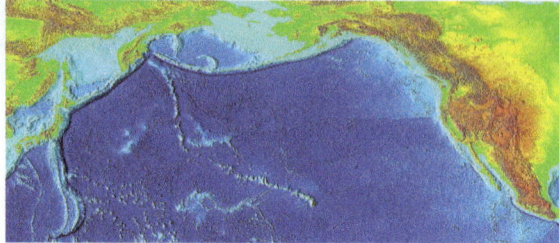

Over millions of years, the Pacific Plate has moved over the Hawaii hotspot, creating a trail of underwater mountains that stretch across the Pacific

Kilauea is the most active shield volcano in the world. The volcano has erupted nonstop since 1983 and it is part of the Hawaiian–Emperor seamount chain.

Mauna Loa is a large shield volcano. Its last eruption was in 1984 and it is part of the Hawaiian–Emperor seamount chain.

Geologists have tried to use hotspot volcanic chains to track the movement of the Earth's tectonic plates. This effort has been vexed by the lack of very long chains, by the fact that many are not time-progressive (e.g. the Galápagos) and by the fact that hotspots do not appear to be fixed relative to one another (e.g. Hawaii and Iceland.)

Postulated Hotspot Volcano Chains

- Hawaiian–Emperor seamount chain (Hawaii hotspot)

- Louisville seamount chain (Louisville hotspot)

- Walvis Ridge (Gough and Tristan hotspot)

- Kodiak–Bowie Seamount chain (Bowie hotspot)

- Cobb–Eickelberg Seamount chain (Cobb hotspot)

- New England Seamount chain (New England hotspot)

- Anahim Volcanic Belt (Anahim hotspot)

- Mackenzie dike swarm (Mackenzie hotspot)

- Great Meteor hotspot track (New England hotspot)

- St. Helena Seamount Chain – Cameroon Volcanic Line (Saint Helena hotspot)

- Southern Mascarene Plateau–Chagos-Maldives-Laccadive Ridge (Réunion hotspot)

- Ninety East Ridge (Kerguelen hotspot)

- Tuamotu–Line Island chain (Easter hotspot)

- Austral–Gilbert–Marshall chain (Macdonald hotspot)

- Juan Fernández Ridge (Juan Fernández hotspot)

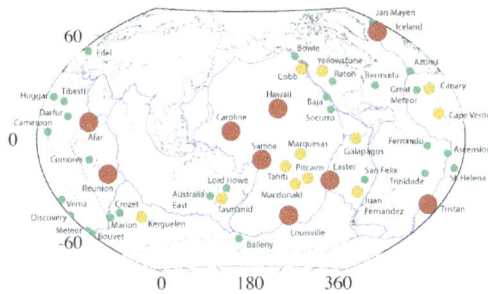

An example of mantle plume locations suggested by one recent group. Figure from Foulger (2010).

List of Volcanic Regions Postulated to be Hotspots

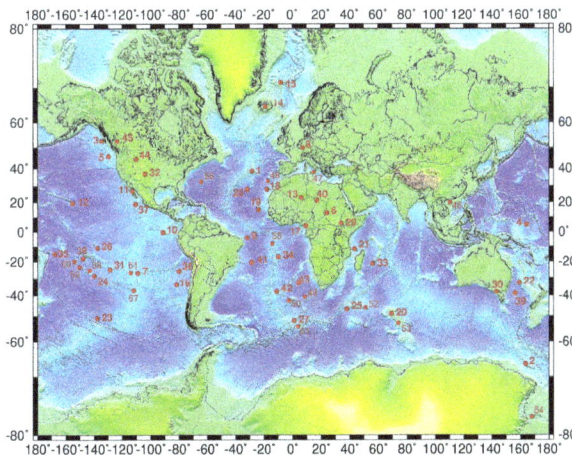

Distribution of hotspots in the list to the left, with the numbers corresponding to those in the list. The Afar hotspot is misplaced.

Map all coordinates using OSM Map all coordinates using Google Map up to 200 coordinates using Bing
Export all coordinates as KML
Export all coordinates as GeoRSS
Export all coordinates as GPX
Map all microformatted coordinates
Place data as RDF

Eurasian Plate

- Eifel hotspot (8)

  - 50°12′N 6°42′E50.2°N 6.7°E, w= 1 az= 082° ±8° rate= 12 ±2 mm/yr

- Iceland hotspot (14)

  - 64°24′N 17°18′W64.4°N 17.3°W

    - Eurasian Plate, w= .8 az= 075° ±10° rate= 5 ±3 mm/yr

    - North American Plate, w= .8 az= 287° ±10° rate= 15 ±5 mm/yr

  - Possibly related to the North Atlantic continental rifting (62 Ma), Greenland.

- Azores hotspot (1)

  - 37°54′N 26°00′W37.9°N 26.0°W

    - Eurasian Plate, w= .5 az= 110° ±12°

    - North American Plate, w= .3 az= 280° ±15°

- Jan Mayen hotspot (15)

  - 71°N 9°W71°N 9°W

- Hainan hotspot (46)

  - 20°N 110°E20°N 110°E, az= 000° ±15°

African Plate

- Mount Etna (47)

  - 37°45.304′N 14°59.715′E37.755067°N 14.995250°E

- Hoggar hotspot (13)

  - 23°18′N 5°36′E23.3°N 5.6°E, w= .3 az= 046° ±12°

- Tibesti hotspot (40)
  - 20°48′N 17°30′E20.8°N 17.5°E, w= .2 az= 030° ±15°
- Jebel Marra/Darfur hotspot (6)
  - 13°00′N 24°12′E13.0°N 24.2°E, w= .5 az= 045° ±8°
- Afar hotspot (29, misplaced in map)
  - 7°00′N 39°30′E7.0°N 39.5°E, w= .2 az= 030° ±15° rate= 16 ±8 mm/yr
  - Possibly related to the Afar Triple Junction, 30 Ma.
- Cameroon hotspot (17)
  - 2°00′N 5°06′E2.0°N 5.1°E, w= .3 az= 032° ±3° rate= 15 ±5 mm/yr
- Madeira hotspot (48)
  - 32°36′N 17°18′W32.6°N 17.3°W, w= .3 az= 055° ±15° rate= 8 ±3 mm/yr
- Canary hotspot (18)
  - 28°12′N 18°00′W28.2°N 18.0°W, w= 1 az= 094° ±8° rate= 20 ±4 mm/ yr
- New England/Great Meteor hotspot (28)
  - 29°24′N 29°12′W29.4°N 29.2°W, w= .8 az= 040° ±10°
- Cape Verde hotspot (19)
  - 16°00′N 24°00′W16.0°N 24.0°W, w= .2 az= 060° ±30°
- St. Helena hotspot (34)
  - 16°30′S 9°30′W16.5°S 9.5°W, w= 1 az= 078° ±5° rate= 20 ±3 mm/yr
- Gough hotspot (49), at 40°19' S 9°56' W.
  - 40°18′S 10°00′E40.3°S 10°E, w= .8 az= 079° ±5° rate= 18 ±3 mm/yr
- Tristan hotspot (42), at 37°07' S 12°17' W.
  - 37°12′S 12°18′W37.2°S 12.3°W
- Vema hotspot (Vema Seamount, 43), at 31°38' S 8°20' E.
  - 32°06′S 6°18′W32.1°S 6.3°W
  - Related maybe to the Paraná and Etendeka traps (c. 132 Ma) through the Walvis Ridge.
- Discovery hotspot (50) (Discovery Seamounts)

- o   43°00′S 2°42′W43.0°S 2.7°W, w= 1 az= 068° ±3°
- Bouvet hotspot (51)
  - o   54°24′S 3°24′E54.4°S 3.4°E
- Shona/Meteor hotspot (27)
  - o   51°24′S 1°00′W51.4°S 1.0°W, w= .3 az= 074° ±6°
- Réunion hotspot (33)
  - o   21°12′S 55°42′E21.2°S 55.7°E, w= .8 az= 047° ±10° rate= 40 ±10 mm/yr
  - o   Possibly related to the Deccan Traps (main events: 68.5–66 Ma)
- Comoros hotspot (21)
  - o   11°30′S 43°18′E11.5°S 43.3°E, w= .5 az=118 ±10° rate=35 ±10 mm/yr

Antarctic Plate

- Marion hotspot (25)
  - o   46°54′S 37°36′E46.9°S 37.6°E, w= .5 az= 080° ±12°
- Crozet hotspot (52)
  - o   46°06′S 50°12′E46.1°S 50.2°E, w= .8 az= 109° ±10° rate= 25 ±13 mm/yr
  - o   Possibly related to the Karoo-Ferrar geologic province (183 Ma)
- Kerguelen hotspot (20)
  - o   49°36′S 69°00′E49.6°S 69.0°E, w= .2 az= 050° ±30° rate= 3 ±1 mm/yr
  - o   Île Saint-Paul and Île Amsterdam could be part of the Kerguelen hotspot trail (St. Paul is probably not another hotspot)
  - o   Related maybe to the Kerguelen Plateau (130 Ma)
- Heard hotspot (53)
  - o   53°06′S 73°30′E53.1°S 73.5°E, w= .2 az= 030° ±20°
- Balleny hotspot (2)
  - o   67°36′S 164°48′E67.6°S 164.8°E, w= .2 az= 325° ±7°
- Erebus hotspot (54)

  o   77°30′S 167°12′E77.5°S 167.2°E

South American Plate

- Trindade/Martin Vaz hotspot (41)

  o   20°30′S 28°48′W20.5°S 28.8°W, w= 1 az= 264° ±5°

- Fernando hotspot (9)

  o   3°48′S 32°24′W3.8°S 32.4°W, w= 1 az= 266° ±7°

  o   Possibly related to the Central Atlantic Magmatic Province (c. 200 Ma)

- Ascension hotspot (55)

  o   7°54′S 14°18′W7.9°S 14.3°W

North American Plate

- Bermuda hotspot (56)

  o   32°36′N 64°18′W32.6°N 64.3°W, w= .3 az= 260° ±15°

- Yellowstone hotspot (44)

  o   44°30′N 110°24′W44.5°N 110.4°W, w= .8 az= 235° ±5° rate= 26 ±5 mm/yr

  o   Possibly related to the Columbia River Basalt Group (17–14 Ma).

- Raton hotspot (32)

  o   36°48′N 104°06′W36.8°N 104.1°W, w= 1 az= 240°±4° rate= 30 ±20 mm/yr

- Anahim hotspot (45)

  o   52°54′0″N 123°44′0″W52.90000°N 123.73333°W (Nazko Cone)

Indo-Australian Plate

- Lord Howe hotspot (22)

  o   34°42′S 159°48′E34.7°S 159.8°E, w= .8 az= 351° ±10°

- Tasmanid hotspot (Gascoyne Seamount, 39)

  o   40°24′S 155°30′E40.4°S 155.5°E, w= .8 az= 007° ±5° rate= 63 ±5 mm/yr

- East Australia hotspot (30)

  - 40°48′S 146°00′E40.8°S 146.0°E, w= .3 az= 000° ±15° rate= 65 ±3 mm/yr

Nazca Plate

- Juan Fernández hotspot (16)

  - 33°54′S 81°48′W33.9°S 81.8°W, w= 1 az= 084° ±3° rate= 80 ±20 mm/yr

- San Felix hotspot (36)

  - 26°24′S 80°06′W26.4°S 80.1°W, w= .3 az= 083° ±8°

- Easter hotspot (7)

  - 26°24′S 106°30′W26.4°S 106.5°W, w= 1 az= 087° ±3° rate= 95 ±5 mm/yr

- Galápagos hotspot (10)

  - 0°24′S 91°36′W0.4°S 91.6°W

    - Nazca Plate, w= 1 az= 096° ±5° rate= 55 ±8 mm/yr

    - Cocos Plate, w= .5 az= 045° ±6°

  - Possibly related to the Caribbean large igneous province (main events: 95–88 Ma).

Pacific Plate

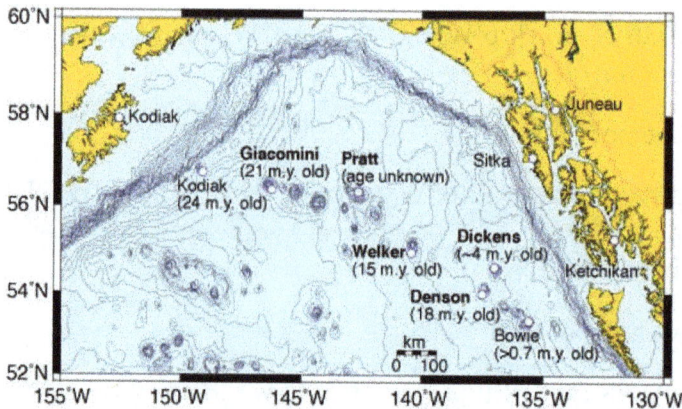

Over millions of years, the Pacific Plate has moved over the Bowie hotspot, creating the Kodiak-Bowie Seamount chain in the Gulf of Alaska

- Louisville hotspot (23)

- 53°36′S 140°36′W53.6°S 140.6°W, w= 1 az= 316° ±5° rate= 67 ±5 mm/yr
  - Possibly related to the Ontong Java Plateau (125–120 Ma).
- Foundation hotspot (57)
  - 37°42′S 111°06′W37.7°S 111.1°W, w= 1 az= 292° ±3° rate= 80 ±6 mm/yr
- Macdonald hotspot (24)
  - 29°00′S 140°18′W29.0°S 140.3°W, w= 1 az= 289° ±6° rate= 105 ±10 mm/yr
- North Austral/President Thiers (President Thiers Bank, 58)
  - 25°36′S 143°18′W25.6°S 143.3°W, w= (1.0) azim= 293° ± 3° rate= 75 ±15 mm/yr
- Arago hotspot (Arago Seamount, 59)
  - 23°24′S 150°42′W23.4°S 150.7°W, w= 1 azim= 296° ±4° rate= 120 ±20 mm/yr
- Maria/Southern Cook hotspot (Îles Maria, 60)
  - 20°12′S 153°48′W20.2°S 153.8°W, w= 0.8 az= 300° ±4°
- Samoa hotspot (35)
  - 14°30′S 168°12′W14.5°S 168.2°W, w= .8 az= 285°±5° rate= 95 ±20 mm/yr
- Crough hotspot (Crough Seamount, 61)
  - 26°54′S 114°36′W26.9°S 114.6°W, w= .8 az= 284° ± 2°
- Pitcairn hotspot (31)
  - 25°24′S 129°18′W25.4°S 129.3°W, w= 1 az= 293° ±3° rate= 90 ±15 mm/yr
- Society/Tahiti hotspot (38)
  - 18°12′S 148°24′W18.2°S 148.4°W, w= .8 az= 295°±5° rate= 109 ±10 mm/yr
- Marquesas hotspot (26)
  - 10°30′S 139°00′W10.5°S 139.0°W, w= .5 az= 319° ±8° rate= 93 ±7 mm/yr
- Caroline hotspot (4)

- - 4°48′N 164°24′E4.8°N 164.4°E, w= 1 az= 289° ±4° rate= 135 ±20 mm/yr

- Hawaii hotspot (12)

  - 19°00′N 155°12′W19.0°N 155.2°W, w= 1 az= 304° ±3° rate= 92 ±3 mm/yr

  - Possibly related to the Siberian Traps (251–250 Ma).

- Socorro/Revillagigedos hotspot (37)

  - 19°00′N 111°00′W19.0°N 111°W

- Guadalupe hotspot (11)

  - 27°42′N 114°30′W27.7°N 114.5°W, w= .8 az= 292° ±5° rate= 80 ±10 mm/yr

- Cobb hotspot (5)

  - 46°00′N 130°06′W46.0°N 130.1°W, w= 1 az= 321° ±5° rate= 43 ±3 mm/yr

- Bowie/Pratt-Welker hotspot (3)

  - 53°00′N 134°48′W53.0°N 134.8°W, w=.8 az= 306° ±4° rate= 40 ±20 mm/yr

Low-velocity Zone

The low-velocity zone (LVZ) occurs close to the boundary between the lithosphere and the asthenosphere in the upper mantle. It is characterized by unusually low seismic shear wave velocity compared to the surrounding depth intervals. This range of depths also corresponds to anomalously high electrical conductivity.It is present between about 80 and 300 km depth. This appears to be universally present for S waves, but may be absent in certain regions for P waves. A second low-velocity zone (not generally referred to as the LVZ, but as ULVZ) has been detected in a thin ≈50 km layer at the core-mantle boundary. These LVZs may have important implications for plate tectonics and the origin of the Earth's crust.

The LVZ has been interpreted to indicate the presence of a significant degree of partial melting, and alternatively as a natural consequence of a thermal boundary layer and the effects of pressure and temperature on the elastic wave velocity of mantle components in the solid state. In any event, a very limited amount of melt (about 1%) is needed to produce these effects. Water in this layer can lower the melting point, and may play an important part in its composition.

Identification

The existence of the low-velocity zone was first proposed from the observation of slower than expected seismic wave arrivals from earthquakes in 1959 by Beno Gutenberg. He noted that between 1° to 15° from the epicenter the longitudinal arrivals showed an exponential decrease in amplitude after which they showed a sudden large increase. The presence of a low-velocity layer that defocussed the seismic energy, followed by a high velocity gradient that concentrated it, provided an explanation for these observations.

Characteristics

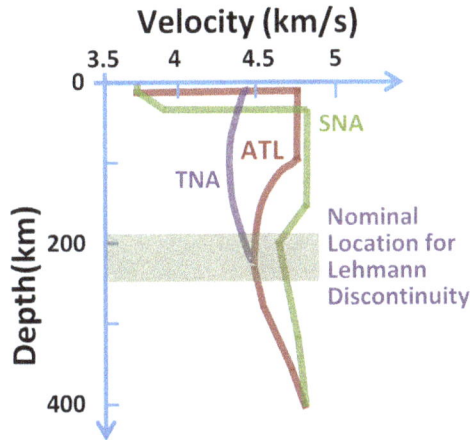

Velocity of seismic S-waves in the Earth near the surface in three tectonic provinces: TNA= Tectonic North America SNA= Shield North America & ATL = North Atlantic.

The LVZ shows a reduction in velocity of about 3–6% with the effect being more pronounced with S-waves compared to P-waves. As is evident from the figure, the reduction and depth over which reduction occurs varies with the choice of tectonic province, that is, regions differ in their seismic characteristics. Following the drop, the base of the zone is marked by an increase in velocity, but it has not been possible to decide whether this transition is sharp or gradual. This lower boundary, found beneath the continental lithosphere and oceanic lithosphere away from mid-ocean ridges, is sometimes referred to as the Lehmann discontinuity and occurs at about 220±30 km depth. The interval also shows a reduction in Q, the seismic quality factor (representing a relatively high degree of seismic attenuation), and a relatively high electrical conductivity.

The LVZ is present at the base of the lithosphere except in areas of thick continental shield where no velocity anomaly is apparent.

Interpretation

The interpretation of these observations is complicated by the effects of seismic anisotropy, which may greatly reduce the actual scale of the velocity anomaly. However, because of the reductions in Q and electrical resistivity in the LVZ, it is generally inter-preted as a zone in which there is a small degree of partial melting. For this to occur at the depths where the LVZ is observed, small amounts of water and/or carbon dioxide must be present to depress the melting point of the silicate minerals. Only 0.05–0.1 % water would be sufficient to cause the 1% of melting necessary to produce the observed changes in physical properties. The lack of LVZ beneath continental shields is explained by the much lower geothermal gradient, preventing any degree of partial melting.

Mantle Plume

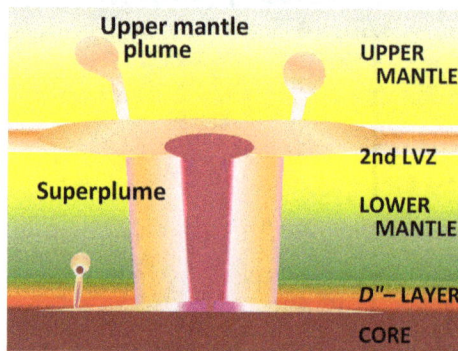

A superplume generated by cooling processes in the mantle (LVZ=Low-velocity zone)

A mantle plume is a mechanism proposed in 1971 to explain volcanic regions of the Earth that were not thought to be explicable by the then-new theory of plate tectonics. Some such volcanic regions lie far from tectonic plate boundaries, for example, Hawaii. Others represent unusually large-volume volcanism, whether on plate boundaries, e.g. Iceland, or basalt floods such as the Deccan or Siberian traps.

A mantle plume is posited to exist where hot rock nucleates at the core-mantle bound-ary and rises through the Earth's mantle becoming a diapir in the Earth's crust. The currently active volcanic centers are known as "hot spots". In particular, the concept that mantle plumes are fixed relative to one another, and anchored at the core-mantle boundary, was thought to provide a natural explanation for the time-progressive chains of older volcanoes seen extending out from some such hot spots, such as the Hawaiian–Emperor seamount chain.

The hypothesis of mantle plumes from depth is not universally accepted as explain-ing all such volcanism. It has required progressive hypothesis-elaboration leading to variant propositions such as mini-plumes and pulsing plumes. Another hypothesis for unusual volcanic regions is the "Plate model". This proposes shallower, passive leakage of magma from the mantle onto the Earth's surface where extension of the lithosphere permits it, attributing most volcanism to plate tectonic processes, with volcanoes far from plate boundaries resulting from intraplate extension.

Concepts

In 1971, geophysicist W. Jason Morgan proposed the hypothesis of mantle plumes. In this hypothesis, convection in the mantle transports heat from the core to the Earth's surface in thermal diapirs. In this concept, two largely independent convective processes occur in the mantle: the broad convective flow associated with *plate tectonics*, which is driven primarily by the sinking of cold plates of lithosphere back into the mantle asthenosphere, and *mantle plumes*, which carry heat upward in narrow, rising columns, driven by heat exchange across the core-mantle boundary. The latter type of convection is postulated to be independent of plate motions.

The sizes and occurrence of mushroom mantle plumes can be predicted easily by transient instability theory developed by Tan and Thorpe. The theory predicts mushroom mantle plumes of about 2000 km diameter with a critical time of about 830 Myr for a core mantle heat flux of 20 mW/m², while the cycle time is about 2 Gyr. The number of mantle plumes is predicted to be about 17.

The plume hypothesis was studied using laboratory experiments conducted in small fluid-filled tanks in the early 1970s. Thermal or compositional fluid-dynamical plumes produced in that way were presented as models for the much larger postulated mantle plumes. On the basis of these experiments, mantle plumes are now postulated to comprise two parts: a long thin conduit connecting the top of the plume to its base, and a bulbous head that expands in size as the plume rises. The entire structure is considered to resemble a mushroom. The bulbous head of thermal plumes forms because hot material moves upward through the conduit faster than the plume itself rises through its surroundings. In the late 1980s and early 1990s, experiments with thermal models showed that as the bulbous head expands it may entrain some of the adjacent mantle into the head.

When a plume head encounters the base of the lithosphere, it is expected to flatten out against this barrier and to undergo widespread decompression melting to form large volumes of basalt magma. It may then erupt onto the surface. Numerical modelling predicts that melting and eruption will take place over several million years. These eruptions have been linked to flood basalts, although many of those erupt over much shorter time scales (less than 1 million years). Examples include the Deccan traps in India, the Siberian traps of Asia, the Karoo-Ferrar basalts/dolerites in South Africa and Antarctica, the Paraná and Etendeka traps in South America and Africa (formerly a single province separated by opening of the South Atlantic Ocean), and the Columbia River basalts of North America. Flood basalts in the oceans are known as oceanic plateaus, and include the Ontong Java plateau of the western Pacific Ocean and the Kerguelen Plateau of the Indian Ocean.

The narrow vertical pipe, or conduit, postulated to connect the plume head to the core-mantle boundary, is viewed as providing a continuous supply of magma to a fixed location, often referred to as a "hot spot". As the overlying tectonic plate (lithosphere) moves over this "hot spot", the eruption of magma from the fixed conduit onto the surface is expected to form a chain of volcanoes that parallels plate motion. The Hawaiian Islands chain in the Pacific Ocean is the type example. Interestingly, it has recently been discovered that the volcanic locus of this chain has not been fixed over time, and it thus joined the club of the many type examples that do not exhibit the key characteristic originally proposed.

The eruption of continental flood basalts is often associated with continental rifting and breakup. This has led to the hypothesis that mantle plumes contribute to continental rifting and the formation of ocean basins. In the context of the alternative "Plate model", continental breakup is a process integral to plate tectonics, and massive volcanism occurs as a natural consequence when it onsets.

The current mantle plume theory is that material and energy from Earth's interior are exchanged with the surface crust in two distinct modes: the predominant, steady state plate tectonic regime driven by upper mantle convection, and a punctuated, intermittently dominant, mantle overturn regime driven by plume convection. This second regime, while often discontinuous, is periodically significant in mountain building and continental breakup.

Chemistry, Heat Flow and Melting

The chemical and isotopic composition of basalts found at "hot spots" differs subtly from mid-ocean-ridge basalts. This geochemical signature arises from the mixing of near-surface materials such as subducted slabs and continental sediments, in the mantle source. There are two competing interpretations for this. In the context of mantle plumes, the near-surface material is postulated to have been transported down to the core-mantle boundary by subducting slabs, and to have been transported back up to the surface by plumes. In the context of the Plate hypothesis, subducted material is mostly re-circulated in the shallow mantle and tapped from there by volcanoes.

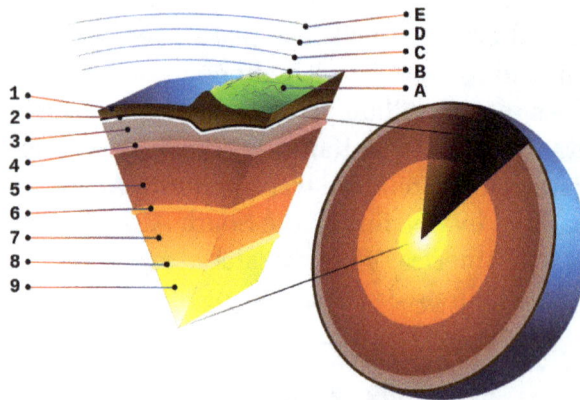

Earth cross-section showing location of upper (3) and lower (5) mantle, D''-layer (6), and outer (7) and inner (9) core

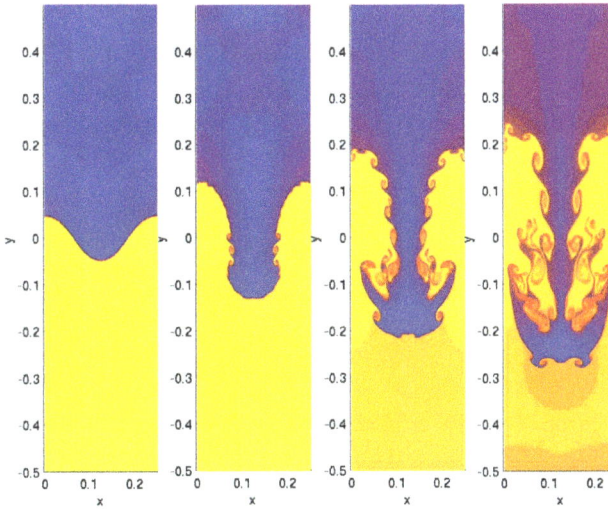

Hydrodynamic simulation of a single "finger" of the Rayleigh–Taylor instability, a possible mechanism for plume formation. In the third and fourth frame in the sequence, the plume forms a "mushroom cap". Note that the core is at the top of the diagram and the crust is at the bottom.

The processing of oceanic crust, lithosphere, and sediment through a subduction zone decouples the water-soluble trace elements (e.g., K, Rb, Th) from the immobile trace elements (e.g., Ti, Nb, Ta), concentrating the immobile elements in the oceanic slab (the water-soluble elements are added to the crust in island arc volcanoes). Seismic tomography shows that subducted oceanic slabs sink as far as the bottom of the mantle transition zone at 650 km depth. Subduction to greater depths is less certain, but there is evidence that they may sink to mid-lower-mantle depths at about 1,500 km depth.

The source of mantle plumes, is postulated to be the core-mantle boundary at 3,000 km depth. Because there is little material transport across the core-mantle boundary, heat transfer must occur by conduction, with adiabatic gradients above and below this boundary. The core-mantle boundary is a strong thermal (temperature) discontinuity. The temperature of the core is approximately 1,000 degrees Celsius higher than that of the overlying mantle. Plumes are postulated to rise as the base of the mantle becomes hotter and more buoyant.

Plumes are postulated to rise through the mantle and begin to partially melt on reaching shallow depths in the asthenosphere by decompression melting. This would create large volumes of magma. The plume hypothesis postulates that this melt rises to the surface and erupts to form "hot spots".

The Lower Mantle and the Core

The most prominent thermal contrast known to exist in the deep (1000 km) mantle is at the core-mantle boundary. Mantle plumes were originally postulated to rise from this layer because the "hot spots" that are assumed to be their surface expression were thought to be fixed relative to one another. This required that plumes were sourced from beneath the shallow asthenosphere that is thought to be flowing rapidly in response to motion of the overlying tectonic plates. There is no other known major thermal boundary layer in the deep Earth, and so the core-mantle boundary was the only candidate.

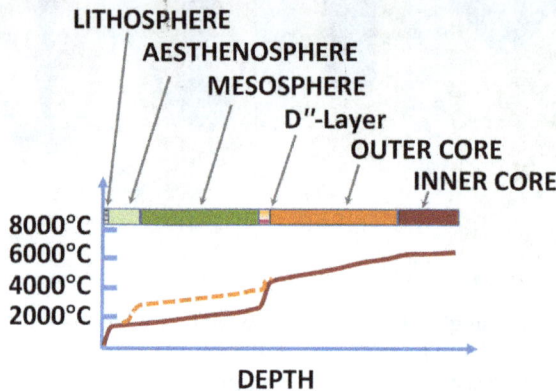

Calculated Earth's temperature vs. depth. Dashed curve: Layered mantle convection; Solid curve: Whole mantle convection.

The base of the mantle is known as the D″ layer, a seismological subdivision of the Earth. It appears to be compositionally distinct from the overlying mantle, and may contain partial melt.

Two very large, broad, large low-shear-velocity provinces, exist in the lower mantle under Africa and under the central Pacific. It is postulated that small plumes rise from their surface or their edges. Their low seismic velocities were thought to suggest that they are relatively hot, although it has recently been shown that their low wave velocities are due to high density caused by chemical heterogeneity.

Evidence for the Theory

Various lines of evidence have been cited in support of mantle plumes. There is some confusion regarding what constitutes support, as there has been a tendency to re-define the postulated characteristics of mantle plumes after observations have been made.

Some common and basic lines of evidence cited in support of the theory are linear volcanic chains, noble gases, geophysical anomalies and geochemistry.

Linear Volcanic Chains

The age-progressive distribution of the Hawaiian-Emperor seamount chain has been explained as a result of a fixed, deep-mantle plume rising into the upper mantle, partly melting, and causing a volcanic chain to form as the plate moves overhead relative to the fixed plume source. Other "hot spots" with time-progressive volcanic chains behind them include Réunion and the Laccadives-Chagos Ridge, the Louisville seamount chain, the Ninety East Ridge and Kerguelen, Tristan da Cunha, and Yellowstone.

An intrinsic aspect of the plume hypothesis is that the "hot spots" and their volcanic trails have been fixed relative to one another throughout geological time. Whereas there is evidence that the chains listed above are time-progressive, it has, however, been shown that they are not fixed relative to one another. The most remarkable example of this is the Emperor chain, the older part of the Hawaii system, which was formed by migration of volcanic activity across a geo-stationary plate.

Many postulated "hot spots" are also lacking time-progressive volcanic trails, e.g., Iceland, the Galapagos, and the Azores. Mismatches between the predictions of the hypothesis and observations are commonly explained by auxiliary processes such as "mantle wind", "ridge capture", "ridge escape" and lateral flow of plume material.

Noble Gas and other Isotopes

^{3}He is considered a primordial isotope as it was formed in the Big Bang. Very little is produced, and little has been added to the Earth by other processes since then. ^{4}He includes a primordial component, but it is also produced by the natural radioactive decay of U and Th. Over time, He in the upper atmosphere is lost into space. Thus, the Earth has become progressively depleted in He, and ^{3}He is not replaced as ^{4}He is. As a result, the ratio ^{3}He/^{4}He in the Earth has lowered over time.

Unusually high ^{3}He/^{4}He have been observed in some, but not all, "hot spots". In mantle plume theory, this is explained by plumes tapping a deep, primordial reservoir in the lower mantle, where the original, high ^{3}He/^{4}He ratios have been preserved throughout geologic time. In the context of the Plate hypothesis, the high ratios are explained by preservation of old material in the shallow mantle. Ancient, high ^{3}He/^{4}He ratios would be particularly easily preserved in materials lacking U or Th, so ^{4}He was not added over time. Olivine and dunite, both found in subducted crust, are materials of this sort.

Other elements, e.g. osmium, have been suggested to be tracers of material arising from near to the Earth's core, in basalts at oceanic islands. However, so far conclusive proof for this is lacking.

Geophysical Anomalies

The plume hypothesis has been tested by looking for the geophysical anomalies predicted to be associated with them. These include thermal, seismic, and elevation anomalies. Thermal anomalies are inherent in the term "hot spot". They can be measured in numerous different ways, including surface heat flow, petrology, and seismology. Thermal anomalies produce anomalies in the speeds of seismic waves, but unfortunately so do composition and partial melt. As a result, wave speeds cannot be used simply and directly to measure temperature, but more sophisticated approaches must be taken.

Diagram showing a cross section though the Earth's lithosphere (in yellow) with magma rising from the mantle (in red). The crust may translate relative to the plume, creating a *track*.

Seismic anomalies are identified by mapping variations in wave speed as seismic waves travel through Earth. A hot mantle plume is predicted to have lower seismic wave speeds compared with similar material at a lower temperature. Mantle material containing a trace of partial melt (e.g., as a result of it having a lower melting point), or being richer in Fe, also has a lower seismic wave speed and those effects are stronger than temperature. Thus, although unusually low wave speeds have been taken to indicate anomalously hot mantle beneath "hot spots", this interpretation is ambiguous. The most commonly cited seismic wave-speed images that are used to look for variations in regions where plumes have been proposed come from seismic tomography. This method involves using a network of seismometers to construct three-dimensional images of the variation in seismic wave speed throughout the mantle.

Seismic waves generated by large earthquakes enable structure below the Earth's surface to be determined along the ray path. Seismic waves that have traveled a thousand or more kilometers (also called teleseismic waves) can be used to image large regions of Earth's mantle. They also have limited resolution, however, and only structures at least several hundred kilometers in diameter can be detected.

Seismic tomography images have been cited as evidence for a number of mantle plumes in Earth's mantle. There is, however, vigorous on-going discussion regarding whether the structures imaged are reliably resolved, and whether they correspond to columns of hot, rising rock.

The mantle plume hypothesis predicts that domal topographic uplifts will develop when plume heads impinge on the base of the lithosphere. An uplift of this kind occurred when the north Atlantic ocean opened about 54 million years ago. Some scientists have linked this to a mantle plume postulated to have caused the breakup of Eurasia and the opening of the north Atlantic, now suggested to underlie Iceland. Current research has shown that the time-history of the uplift is probably much shorter than predicted, however. It is thus not clear how strongly this observation supports the mantle plume hypothesis.

Geochemistry

Basalts found at oceanic islands are geochemically distinct from those found at mid-ocean ridges and volcanoes associated with subduction zones (island arc basalts). "Ocean-island basalt" is also similar to basalts found throughout the oceans on both small and large seamounts (thought to be formed by eruptions on the sea floor that did not rise above the surface of the ocean). They are also compositionally similar to some basalts found in the interiors of the continents (e.g., the Snake River Plain).

In major elements, ocean island basalts are typically higher in iron (Fe) and titanium (Ti) than mid-ocean ridge basalts at similar magnesium (Mg) contents. In trace elements, they are typically more enriched in the light rare earth elements than mid-ocean ridge basalts. Compared to island arc basalts, ocean island basalts are lower in alumina (Al_2O_3) and higher in immobile trace elements (e.g., Ti, Nb, Ta).

These differences result from processes that occur during the subduction of oceanic crust and mantle lithosphere. Oceanic crust (and to a lesser extent, the underlying mantle) typically becomes hydrated to varying degrees on the seafloor, partly as the result of seafloor weathering, and partly in response to hydrothermal circulation near the mid-ocean-ridge crest where it was originally formed. As oceanic crust and underlying lithosphere subduct, water is released by dehydration reactions, along with water-soluble elements and trace elements. This enriched fluid rises to metasomatize the overlying mantle wedge and leads to the formation of island arc basalts. The subducting slab is depleted in these water-mobile elements (e.g., K, Rb, Th, Pb) and thus relatively enriched in elements that are not water-mobile (e.g., Ti, Nb, Ta) compared to both mid-ocean ridge and island arc basalts.

Ocean island basalts are also relatively enriched in immobile elements relative to the water-mobile elements. This, and other observations, have been interpreted as indicating that the distinct geochemical signature of ocean island basalts results from inclusion of a component of subducted slab material. This must have been recycled in the mantle, then re-melted and incorporated in the lavas erupted. In the context of the plume hypothesis, subducted slabs are postulated to have been subducted down as far as the core-mantle boundary, and transported back up to the surface in rising plumes. In the plate hypothesis, the slabs are postulated to have been recycled at shallower depths – in the upper few hundred kilometers that make up the upper mantle. However, the plate hypothesis is inconsistent with both the geochemistry of shallow asthenosphere melts (i.e., Mid-ocean ridge basalts) and with the isotopic compositions of ocean island basalts.

Seismology

In 2015, based on data from 273 large earthquakes, researchers compiled a model based on whole waveform tomography, requiring the equivalent of 3 million hours of supercomputer time. Due to computational limitations, high-frequency data still could not be used, and seismic data remained unavailable from much of the seafloor. Nonetheless, vertical plumes, 400 C hotter than the surrounding rock, were visualized under many hotspots, including the Pitcairn, Macdonald, Samoa, Tahiti, Marquesas, Galapagos, Cape Verde, and Canary hotspots. They extended nearly vertically from the core-mantle boundary (2900 km depth) to a possible layer of shearing and bending at 1000 km. They were detectable because they were 600–800 km wide, more than three times the width expected from contemporary models. Many of these plumes are in the Large Low Shear Velocity Provinces under Africa and the Pacific, while some other hotspots such as Yellowstone were less clearly related to mantle features in the model.

The unexpected size of the plumes leaves open the possibility that they may conduct the bulk of the Earth's 44 terawatts of internal heat flow from the core to the surface, and means that the lower mantle convects less than expected, if at all. It is possible that there is a compositional difference between plumes and the surrounding mantle that slows them down and broadens them.

Suggested Mantle Plume Locations

Many different localities have been suggested to be underlain by mantle plumes, and scientists cannot agree on a definitive list. Some scientists suggest that several tens of plumes exist, whereas others suggest that there are none. The theory was really inspired by the Hawaiian volcano system. Hawaii is a large volcanic edifice in the center of the Pacific ocean, far from any plate boundaries. Its regular, time-progressive chain of islands and seamounts superficially fits the plume theory well. However, it is almost unique on Earth, as nothing as extreme exists anywhere else. The second strongest candidate for a plume location is often quoted to be Iceland, but this lies on a spreading plate boundary, and its massive nature can be equally explained by a combination of plate tectonic forces.

Mantle plumes have been suggested as the source for flood basalts. These extremely rapid, large scale eruptions of basaltic magmas have periodically formed continental flood basalt provinces on land and oceanic plateaus in the ocean basins, such as the Deccan Traps, the Siberian Traps the Karoo/Ferrar flood basalts of Gondwana, and the largest known continental flood basalt, the Central Atlantic magmatic province (CAMP).

Others, have pointed out the coincidence of many continental flood basalt events with continental rifting. This is consistent with a system that tends toward equilibrium, as matter rises in a mantle plume, other material is drawn down into the mantle causing rifting.

Alternative Hypotheses

In parallel with the mantle plume model, two alternative explanations for the observed phenomena have been considered: the plate hypothesis and the impact hypothesis.

The Plate Hypothesis

The plate hypothesis suggests that "anomalous" volcanism results from lithospheric extension that permits melt to rise passively from the asthenosphere beneath. It is thus the conceptual inverse of the plume hypothesis, attributing volcanism to shallow, near-surface processes associated with plate tectonics, rather than active processes arising at the core-mantle boundary. The Plate hypothesis embodies the concept that deep mantle plumes causing surface, time-progressive volcanism do not exist.

Lithospheric extension is attributed to processes related to plate tectonics. These processes are well understood at mid-ocean ridges, where most of Earth's volcanism occurs. It is less commonly recognised that the plates themselves deform internally, and can permit volcanism in those regions where the deformation is extensional. Well-known examples are the Basin and Range Province in the western USA, the East African rift valley, and the Rhine Graben. Variable fertility in the source region, usually the mantle, results in variable volumes of magma being produced. The ocean-island basalt (OIB) geochemistry of lavas found at many places, and attributed to plumes, is, in fact, a geochemical signature of enhanced fertility in the melt source.

The Plate hypothesis thus attributes all of Earth's volcanism to a single process – plate tectonics – rather than to two independent processes (plumes and plate tectonics), but does not address issues of core–mantle heat and/or material transfer.

Under the umbrella of the Plate hypothesis, the following sub-processes, all of which can contribute to permitting surface volcanism, are recognised:

- Continental break-up;
- Fertility at mid-ocean ridges;
- Enhanced volcanism at plate boundary junctions;
- Small-scale sublithospheric convection;
- Oceanic intraplate extension;
- Slab tearing and break-off;
- Shallow mantle convection;
- Abrupt lateral changes in stress at structural discontinuities;
- Continental intraplate extension;

- Catastrophic lithospheric thinning;

- Sublithospheric melt ponding and draining.

The Impact Hypothesis

In addition to these processes, impact events such as ones that created the Addams crater on Venus and the Sudbury Igneous Complex in Canada are known to have caused melting and volcanism. In the impact hypothesis, it is proposed that some regions of hotspot volcanism can be triggered by certain large-body oceanic impacts which are able to penetrate the thinner oceanic lithosphere, and flood basalt volcanism can be triggered by converging seismic energy focused at the antipodal point opposite major impact sites. Impact-induced volcanism has not been adequately studied and comprises a separate causal category of terrestrial volcanism with implications for the study of hotspots and plate tectonics.

Comparison of the Hypotheses

In 1997 it became possible using seismic tomography to image submerging tectonic slabs penetrating from the surface all the way to the core-mantle boundary.

For the Hawaii hotspot, long-period seismic body wave diffraction tomography provided evidence that a mantle plume is responsible, as had been proposed as early as 1971. For the Yellowstone hotspot, seismological evidence began to converge from 2011 in support of the plume model, as concluded by James et al., "we favor a lower mantle plume as the origin for the Yellowstone hotspot." Data acquired through EarthScope, a program collecting high resolution seismic data throughout the conterminous United States has accelerated acceptance of a plume underlying Yellowstone.

Although there is strong evidence that at least two deep mantle plumes rise to the core-mantle boundary, confirmation that other hypotheses can be dismissed may require similar tomographic evidence for other hot spots.

Geologic Modelling

Geological mapping software displaying a screenshot of a structure map generated for an 8500ft deep gas & Oil reservoir in the Erath field, Vermilion Parish, Erath, Louisiana. The left-to-right gap, near the top of the contour map indicates a Fault line. This fault line is between the blue/green contour lines and the purple/red/yellow contour lines. The thin red circular contour line in the middle of the map indicates the top of the oil reservoir. Because gas floats above oil, the thin red contour line marks the gas/oil contact zone.

Geologic modelling, Geological modelling or Geomodelling is the applied science of creating computerized representations of portions of the Earth's crust based on geophysical and geological observations made on and below the Earth surface. A Geomodel is the numerical equivalent of a three-dimensional geological map complemented by a description of physical quantities in the domain of interest. Geomodelling is related to the concept of Shared Earth Model; which is a multidisciplinary, interoperable and updatable knowledge base about the subsurface.

Geomodelling is commonly used for managing natural resources, identifying natural hazards, and quantifying geological processes, with main applications to oil and gas fields, groundwater aquifers and ore deposits. For example, in the oil and gas industry, realistic geologic models are required as input to reservoir simulator programs, which predict the behavior of the rocks under various hydrocarbon recovery scenarios. A reservoir can only be developed and produced once; therefore, making a mistake by selecting a site with poor conditions for development is tragic and wasteful. Using geological models and reservoir simulation allows reservoir engineers to identify which recovery options offer the safest and most economic, efficient, and effective development plan for a particular reservoir.

Geologic modelling is a relatively recent subdiscipline of geology which integrates structural geology, sedimentology, stratigraphy, paleoclimatology, and diagenesis;

In 2-dimensions (2D), a geologic formation or unit is represented by a polygon, which can be bounded by faults, unconformities or by its lateral extent, or crop. In geological models a geological unit is bounded by 3-dimensional (3D) triangulated or gridded surfaces. The equivalent to the mapped polygon is the fully enclosed geological unit, using a triangulated mesh. For the purpose of property or fluid modelling these volumes can be separated further into an array of cells, often referred to as voxels (volumetric elements). These 3D grids are the equivalent to 2D grids used to express properties of single surfaces.

Geomodelling generally involves the following steps:

1. Preliminary analysis of geological context of the domain of study.

2. Interpretation of available data and observations as point sets or polygonal lines (e.g. "fault sticks" corresponding to faults on a vertical seismic section).

3. Construction of a structural model describing the main rock boundaries (horizons, unconformities, intrusions, faults)

4. Definition of a three-dimensional mesh honoring the structural model to support volumetric representation of heterogeneity and solving the Partial Differential Equations which govern physical processes in the subsurface (e.g. seismic wave propagation, fluid transport in porous media).

Geologic Modelling Components

Structural Framework

Incorporating the spatial positions of the major formation boundaries, including the effects of faulting, folding, and erosion (unconformities). The major stratigraphic divisions are further subdivided into layers of cells with differing geometries with relation to the bounding surfaces (parallel to top, parallel to base, proportional). Maximum cell dimensions are dictated by the minimum sizes of the features to be resolved (everyday example: On a digital map of a city, the location of a city park might be adequately resolved by one big green pixel, but to define the locations of the basketball court, the baseball field, and the pool, much smaller pixels - higher resolution - need to be used).

Rock Type

Each cell in the model is assigned a rock type. In a coastal clastic environment, these might be beach sand, high water energy marine upper shoreface sand, intermediate water energy marine lower shoreface sand, and deeper low energy marine silt and shale. The distribution of these rock types within the model is controlled by several methods, including map boundary polygons, rock type probability maps, or statistically emplaced based on sufficiently closely spaced well data.

Reservoir Quality

Reservoir quality parameters almost always include porosity and permeability, but may include measures of clay content, cementation factors, and other factors that affect the storage and deliverability of fluids contained in the pores of those rocks. Geostatistical techniques are most often used to populate the cells with porosity and permeability values that are appropriate for the rock type of each cell.

Fluid Saturation

A 3D finite difference grid used in MODFLOW for simulating groundwater flow in an aquifer.

Most rock is completely saturated with groundwater. Sometimes, under the right conditions, some of the pore space in the rock is occupied by other liquids or gases. In the energy industry, oil and natural gas are the fluids most commonly being modelled. The preferred methods for calculating hydrocarbon saturations in a geologic model incorporate an estimate of pore throat size, the densities of the fluids, and the height of the cell above the water contact, since these factors exert the strongest influence on capillary action, which ultimately controls fluid saturations.

Geostatistics

An important part of geologic modelling is related to geostatistics. In order to represent the observed data, often not on regular grids, we have to use certain interpolation techniques. The most widely used technique is kriging which uses the spatial correlation among data and intends to construct the interpolation via semi-variograms. To reproduce more realistic spatial variability and help assess spatial uncertainty between data, geostatistical simulation based on variograms, training images, or parametric geological objects is often used.

Mineral Deposits

Geologists involved in mining and mineral exploration use geologic modelling to determine the geometry and placement of mineral deposits in the subsurface of the earth. Geologic models help define the volume and concentration of minerals, to which economic constraints are applied to determine the economic value of the mineralization. Mineral deposits that are deemed to be economic may be developed into a mine.

Technology

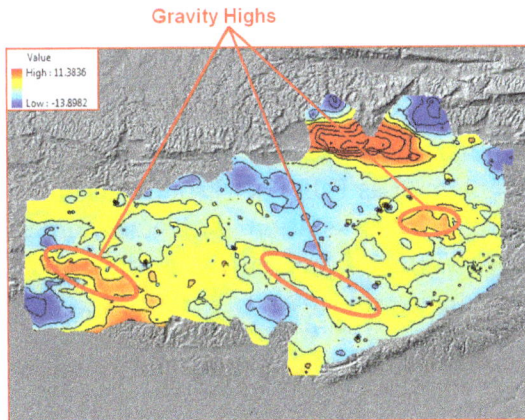

Gravity Highs

Geomodelling and CAD share a lot of common technologies. Software is usually implemented using object-oriented programming technologies in C++, Java or C# on one or multiple computer platforms. The graphical user interface generally consists of one or several 3D and 2D graphics windows to visualize spatial data, interpretations and modelling output. Such visualization is generally achieved by exploiting graphics hardware. User interaction is mostly performed through mouse and key-board, although 3D pointing devices and immersive environments may be used in some specific cases. GIS (Geographic Information System) is also a widely used tool to manipulate geological data.

Geometric objects are represented with parametric curves and surfaces or discrete models such as polygonal meshes.

Research in Geomodelling

Problems pertainting to Geomodelling cover:

- Defining an appropriate Ontology to describe geological objects at various scales of interest,

- Integrating diverse types of observations into 3D geomodels: geological map-ping data, borehole data and interpretations, seismic images and interpreta-tions, potential field data, well test data, etc.,

- Better accounting for geological processes during model building,

- Characterizing uncertainty about the geomodels to help assess risk. Therefore, Geomodelling has a close connection to Geostatistics and Inverse problem the-ory,

- Applying of the recent developed Multiple Point Geostatistical Simulations (MPS) for integrating different data sources,

- Automated geometry optimization and topology conservation

History

In the 70's, geomodelling mainly consisted of automatic 2D cartographic techniques such as contouring, implemented as FORTRAN routines communicating directly with plotting hardware. The advent of workstations with 3D graphics capabilities during the 80's gave birth to a new generation of geomodelling software with graphical user inter-face which became mature during the 90's.

Since its inception, geomodelling has been mainly motivated and supported by oil and gas industry.

Geologic Modelling Software

Software developers have built several packages for geologic modelling purposes. Such software can display, edit, digitise and automatically calculate the parameters required by engineers, geologists and surveyors. Current software is mainly developed and com-mercialized by oil and gas or mining industry software vendors:

Geologic modelling and visualisation

- SGS Genesis
- IRAP RMS Suite
- Geomodeller3D
- Geosoft provides GM-SYS and VOXI 3D modelling software
- GSI3D
- Petrel
- Rockworks
- Move

Groundwater modelling

- FEFLOW
- FEHM
- MODFLOW
  - GMS
  - Visual MODFLOW
- ZOOMQ3D

Moreover, industry Consortia or companies are specifically working at improving standardization and interoperability of earth science databases and geomodelling software:

- Standardization: GeoSciML by the Commission for the Management and Application of Geoscience Information, of the International Union of Geological Sciences.
- Standardization: RESQML(tm) by Energistics
- Interoperability: OpenSpirit, by TIBCO(r)

Synthetic Seismogram

A synthetic seismogram is the result of forward modelling the seismic response of an input earth model, which is defined in terms of 1D, 2D or 3D variations in physical properties. In hydrocarbon exploration this is used to provide a 'tie' between changes in rock properties in a borehole and seismic reflection data at the same location. It can also be used either to test possible interpretation models for 2D and 3D seismic data or to model the response of the predicted geology as an aid to planning a seismic reflection survey. In the processing of wide-angle reflection and refraction (WARR) data, synthetic seismograms are used to further constrain the results of seismic tomography. In earthquake seismology, synthetic seismograms are used either to match the predicted effects of a particular earthquake source fault model with observed seismometer records or to help constrain the Earth's velocity structure. Synthetic seismograms are generated using specialized geophysical software.

1D Synthetics

Seismic reflection data are initially only available in the time domain. In order that the geology encountered in a borehole can be tied to the seismic data, a 1D synthetic seismogram is generated. This is important in identifying the origin of seismic reflections seen on the seismic data. Density and velocity data are routinely measured down the borehole using wireline logging tools. These logs provide data with a sampling interval much smaller than the vertical resolution of the seismic data. The logs are therefore often averaged over intervals to produce what is known as a 'blocked-log'. This information is then used to calculate the variation in acoustic impedance down the well bore using the Zoeppritz equations. This acoustic impedance log is combined with the velocity data to generate a reflection coefficient series in time. This series is convolved with a seismic wavelet to produce the synthetic seismogram. The input seismic wavelet is chosen to match as closely as possible to that produced during the original seismic acquisition, paying particular attention to phase and frequency content.

1.5D Seismic Modelling

The convolutional 1D modelling produces seismograms containing approximations of primary reflections only. For more accurate modelling involving multiple reflections, head waves, guided waves and surface waves, as well as transmission effects and geometrical spreading, full waveform modelling is required. For 1D elastic models the most accurate approach to full waveform modelling is known as the reflectivity method. This method is based on the integral transform approach, whereby the wave field (cylinidrical or spherical wave) is represented by a sum (integral) of time-harmonic plane waves. The reflection and transmission coefficients for individual plane waves propagating in a stack of layers can be computed analytically using a variety of methods, such as matrix propagator, global matrix or invariant embedding. This group of methods is called 1.5D because the earth is represented by a 1D model (flat layers), while wave propagation is considered either in 2D (cylindrical waves) or 3D (spherical waves).

2D Synthetic Seismic Modeling

A similar approach can be used to examine the seismic response of a 2D geological cross-section. This can be used to look at such things as the resolution of thin beds or the different responses of various fluids, e.g. oil, gas or brine in a potential reservoir sand. It may also be used to test out different geometries of structures such as salt diapirs, to see which gives the best match to the original seismic data. A cross-section is built with density and seismic velocities assigned to each of the individual layers. These can be either constant within a layer or varying in a systematic fashion across the model both horizontally and vertically. The software program then runs a synthetic acquisition across the model to produce a set of 'shot gathers' that can be processed as if they were real seismic data to produce a synthetic 2D seismic section. The synthetic record is generated using either a ray-tracing algorithm or some form of full waveform modelling, depending on the purpose of the modelling. Ray-tracing is quick and sufficient for testing the illumination of the structure, but full waveform modelling will be necessary to accurately model the amplitude response.

3D Synthetic Seismic Modelling

The approach can be further expanded to model the response of a 3D geological model. This is used to reduce the uncertainty in interpretation by modelling the response of the 3D model to a synthetic seismic acquisition that matches as closely as possible to that actually used in acquiring the data that has been interpreted. The synthetic seismic data is then processed using the same sequence as that used for the original data. This method can be used to model both 2D and 3D seismic data that has been acquired over the area of the geological model. During the planning of a seismic survey, 3D modelling can be used to test the effect of variation in seismic acquisition parameters, such as the shooting direction or the maximum offset between source and receiver, on the imaging of a particular geological structure.

WARR Data Modelling

Wide Aperture Reflection and Refraction (WARR) models' initial processing is normally carried out using a tomographic approach in which the time of observed first arrivals is matched by varying the velocity structure of the subsurface. The model can be further refined using forward modelling to generate synthetic seismograms for individual shot gathers.

Earthquake Modelling

Source Modelling

In areas that have a well understood velocity structure it is possible to use synthetic seismograms to test out the estimated source parameters of an earthquake. Parameters such as the fault plane, slip vector and rupture velocity can be varied to produce synthetic seismic responses at individual seismometers for comparison with the observed seismograms.

Velocity Modelling

For seismic events of known type and location, it is possible to obtain detailed information about the Earth's structure, at various scales, by modelling the teleseismic response of the event.

Seismic Gap

A seismic gap is a segment of an active fault known to produce significant earthquakes that has not slipped in an unusually long time, compared with other segments along the same structure. There is a hypothesis or theory that states that over long periods of time, the displacement on any segment must be equal to that experienced by all the other parts of the fault. Any large and longstanding gap is, therefore, considered to be the fault segment most likely to suffer future earthquakes.

The applicability of this approach has been criticised by some seismologists although earthquakes sometimes have occurred in previously-identified seismic gaps.

Examples

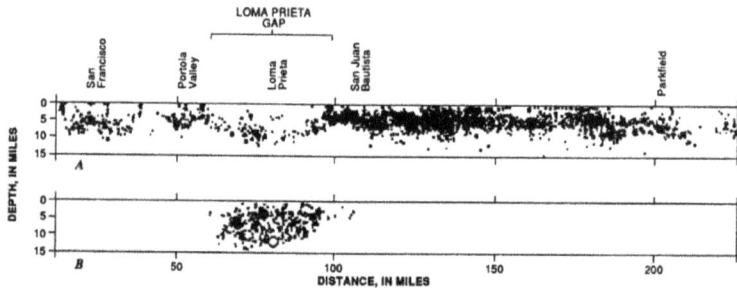

Cross sections along the San Andreas fault showing recorded seismic activity A) 20 years before the Loma Prieta event, B) The main shock (large circle) and aftershocks for the Loma Prieta event, USGS Circular 1045

Loma Prieta Seismic Gap, California

Prior to the 1989 Loma Prieta earthquake, that segment of the San Andreas fault system recorded much less seismic activity than other parts of the fault. The main shock and aftershocks of the 1989 event occurred within the previous seismic gap.

Central Kuril Gap, Russia

Immediately following the 2004 Indian Ocean earthquake, a seismic gap analysis of the seismic zones around the Pacific Ocean identified the Central Kuril segment of the Kuril-Kamchatka Trench subduction zone as the most likely to give rise to a major earthquake. This zone, 500 km in length, at that time had experienced no major earthquake since 1780, but was bounded to north and south by segments that had moved within the last 100 years. The M_w = 8.3 earthquake of 15 November 2006 and the M_w = 8.2 earthquake of 13 January 2007 occurred within the defined gap.

References

- Peter M. Shearer (2009). Introduction to Seismology. Cambridge University Press. ISBN 978-0-521-88210-1.

- Seth Stein; Michael Wysession (1 April 2009). An Introduction to Seismology, Earthquakes, and Earth Structure. John Wiley & Sons. ISBN 978-14443-1131-0.

- Sheriff, R. E., Geldart, L. P. (1995). Exploration Seismology (2nd ed.). Cambridge University Press. p. 52. ISBN 0-521-46826-4.

- Milsom, J. (2003). Field Geophysics. The geological field guide series. 25. John Wiley and Sons. p. 232. ISBN 978-0-470-84347-5. Retrieved 2010-02-25.

- Telford, William Murray; Geldart, L. P.; Robert E. Sheriff (1990). Applied geophysics. Cambridge University Press. p. 149. ISBN 978-0-521-33938-4. Retrieved 8 June 2011.

- L. B. Freund (1998). Dynamic Fracture Mechanics. Cambridge University Press. p. 83. ISBN 978-0521629225.

- Thompson, Donald O.; Chimenti, Dale E. (1 June 1997). Review of progress in quantitative nondestructive evaluation. Springer. p. 161. ISBN 978-0-306-45597-1. Retrieved 8 June 2011.

- Biryukov, S.V.; Gulyaev, Y.V.; Krylov, V.V.; Plessky, V.P. (1995). Surface Acoustic Waves in Inhomogeneous Media. Springer. ISBN 978-3-642-57767-3.

- Fossen, H. (2010). Structural Geology. Cambridge University Press. ISBN 9780521516648. Retrieved 27 January 2013.

- Kearey, Philip; Klepeis, Keith A.; Vine, Frederick J. (2013-05-28). Global Tectonics. John Wiley & Sons. ISBN 1118688082.

- Foulger, G.R. (2010). Plates vs. Plumes: A Geological Controversy. Wiley-Blackwell. ISBN 978-1-4051-6148-0.

- Philip Kearey; Keith A. Klepeis; Frederick J. Vine (2009). Global tectonics (3rd ed.). Wiley-Blackwell. p. 32. ISBN 1-4051-0777-4.

- Gutenberg, B. (1959). Physics of the Earth's Interior. New York: Academic Press. p. 240. ISBN 0-12-310650-8.

- Anderson, D.L. (1989). "3. The Crust and Upper Mantle". Theory of the Earth (PDF). Boston: Blackwell Scientific Publications. ISBN 0-521-84959-4. Retrieved 2010-02-20.

- Figure patterned after Don L Anderson (2007). New theory of the earth (2nd ed.). Cambridge University Press. p. 102, Figure 8.6. ISBN 0-521-84959-4.

- Brown, G.C.; Mussett A.E. (1981). The inaccessible earth. Taylor & Francis. p. 235. ISBN 978-0-04-550028-4. Retrieved 2010-02-20.

- Condie, K.C. (1997). Plate tectonics and crustal evolution. Butterworth-Heinemann. p. 282. ISBN 978-0-7506-3386-4. Retrieved 2010-02-20.

- Foulger, Gillian R. (2005). Plates, plumes, and paradigms; Volume 388 of Special papers.

Geological Society of America. p. 195. ISBN 0-8137-2388-4.

- Condie, Kent C. (1997). Plate tectonics and crustal evolution (4th ed.). Butterworth-Heinemann. p. 5. ISBN 0-7506-3386-7.

- "Seismic Tomography—Using earthquakes to image Earth's interior". Incorporated Research Institutions for Seismology (IRIS). Retrieved 18 May 2016.

- Julian, Brian (2006). "Seismology: The Hunt for Plumes" (PDF). mantleplumes.org. Retrieved 3 May 2016.

- "Seismic Tomography" (PDF). earthscope.org. Incorporated Research Institutions for Seismology (IRIS). Retrieved 18 May 2016.

- Dziewonski, Adam. "Global Seismic Tomography: What we really can say and what we make up" (PDF). mantleplumes.org. Retrieved 18 May 2016.

Reflection Seismology: An Overview

Reflection seismology is the method that uses the principles of seismology to assess the substratum after a reflected seismic wave. Seismic migration is the transferring of seismic events because of the complexity in the initial geographical area. The following text provides the reader with an in-depth understanding of reflection seismology.

Reflection Seismology

Reflection seismology (or seismic reflection) is a method of exploration geophysics that uses the principles of seismology to estimate the properties of the Earth's subsurface from reflected seismic waves. The method requires a controlled seismic source of energy, such as dynamite/Tovex, a specialized air gun or a seismic vibrator, commonly known by the trademark name Vibroseis. Reflection seismology is similar to sonar and echolocation.

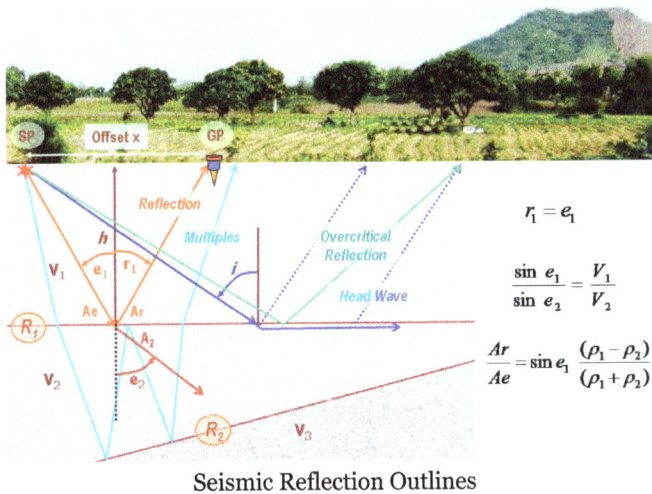

$$r_1 = e_1$$

$$\frac{\sin e_1}{\sin e_2} = \frac{V_1}{V_2}$$

$$\frac{Ar}{Ae} = \sin e_1 \frac{(\rho_1 - \rho_2)}{(\rho_1 + \rho_2)}$$

Seismic Reflection Outlines

Outline of the Method

Seismic waves are mechanical perturbations that travel in the Earth at a speed governed by the acoustic impedance of the medium in which they are travelling. The acoustic (or seismic) impedance, Z, is defined by the equation:

$$Z = V\rho,,$$

where V is the seismic wave velocity and ρ is the density of the rock.

When a seismic wave travelling through the Earth encounters an interface between two materials with different acoustic impedances, some of the wave energy will reflect off the interface and some will refract through the interface. At its most basic, the seismic reflection technique consists of generating seismic waves and measuring the time taken for the waves to travel from the source, reflect off an interface and be detected by an array of receivers (or geophones) at the surface. Knowing the travel times from the source to various receivers, and the velocity of the seismic waves, a geophysicist then attempts to reconstruct the pathways of the waves in order to build up an image of the subsurface.

In common with other geophysical methods, reflection seismology may be seen as a type of inverse problem. That is, given a set of data collected by experimentation and the physical laws that apply to the experiment, the experimenter wishes to develop an abstract model of the physical system being studied. In the case of reflection seismology, the experimental data are recorded seismograms, and the desired result is a model of the structure and physical properties of the Earth's crust. In common with other types of inverse problems, the results obtained from reflection seismology are usually not unique (more than one model adequately fits the data) and may be sensitive to relatively small errors in data collection, processing, or analysis. For these reasons, great care must be taken when interpreting the results of a reflection seismic survey.

The Reflection Experiment

The general principle of seismic reflection is to send elastic waves (using an energy source such as dynamite explosion or Vibroseis) into the Earth, where each layer within the Earth reflects a portion of the wave's energy back and allows the rest to refract through. These reflected energy waves are recorded over a predetermined time period (called the record length) by receivers that detect the motion of the ground in which they are placed. On land, the typical receiver used is a small, portable instrument known as a geophone, which converts ground motion into an analogue electrical signal. In water, hydrophones are used, which convert pressure changes into electrical signals. Each receiver's response to a single shot is known as a "trace" and is recorded onto a data storage device, then the shot location is moved along and the process is repeated. Typically, the recorded signals are subjected to significant amounts of signal processing before they are ready to be interpreted and this is an area of significant active research within industry and academia. In general, the more complex the geology of the area under study, the more sophisticated are the techniques required to remove noise and increase resolution. Modern seismic reflection surveys contain large amount of data and so require large amounts of computer processing, often performed on supercomputers or computer clusters.

Reflection and Transmission at Normal Incidence

When a seismic wave encounters a boundary between two materials with different acoustic impedances, some of the energy in the wave will be reflected at the boundary, while some of the energy will be transmitted through the boundary. The amplitude of the reflected wave is predicted by multiplying the amplitude of the incident wave by the seismic *reflection coefficient* R, determined by the impedance contrast between the two materials.

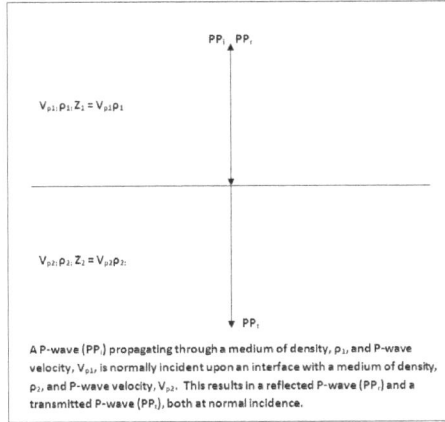

A P-wave (PP_i) propagating through a medium of density, ρ_1, and P-wave velocity, V_{p1}, is normally incident upon an interface with a medium of density, ρ_2, and P-wave velocity, V_{p2}. This results in a reflected P-wave (PP_r) and a transmitted P-wave (PP_t), both at normal incidence.

P-wave reflects off an interface at normal incidence

For a wave that hits a boundary at normal incidence (head-on), the expression for the reflection coefficient is simply

$$R = \frac{Z_2 - Z_1}{Z_2 + Z_1},$$

where Z_1 and Z_2 are the impedance of the first and second medium, respectively.

Similarly, the amplitude of the incident wave is multiplied by the *transmission coefficient* to predict the amplitude of the wave transmitted through the boundary. The formula for the normal-incidence transmission coefficient is

$$T = 1 - R = \frac{2Z_1}{(Z_2 + Z_1)}.$$

As the sum of the squares of amplitudes of the reflected and transmitted wave has to be equal to the square of amplitude of the incident wave, it is easy to show that

$$Z_1(1 - R^2) = \frac{Z_1(Z_2 + Z_1)^2 - Z_1(Z_2 - Z_1)^2}{(Z_2 + Z_1)^2} = \frac{4Z_2 Z_1^2}{(Z_2 + Z_1)^2} = Z_2 T^2.$$

By observing changes in the strength of reflectors, seismologists can infer changes in the seismic impedances. In turn, they use this information to infer changes in the properties of the rocks at the interface, such as density and elastic modulus.

Reflection and Transmission at Non-normal Incidence

The situation becomes much more complicated in the case of non-normal incidence, due to mode conversion between P-waves and S-waves, and is described by the Zoeppritz equations. In 1919, Karl Zoeppritz derived 4 equations that determine the amplitudes of reflected and refracted waves at a planar interface for an incident P-wave as a function of the angle of incidence and six independent elastic parameters. These equations have 4 unknowns and can be solved but they do not give an intuitive understanding for how the reflection amplitudes vary with the rock properties involved.

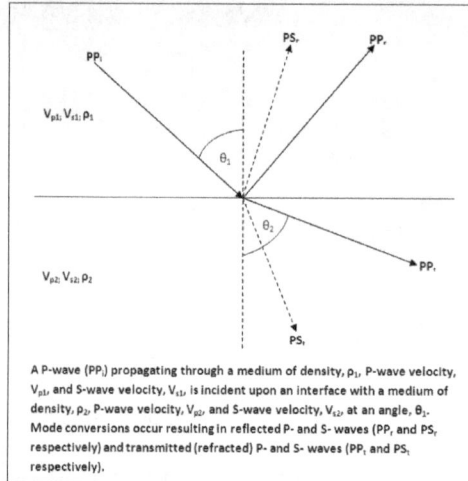

A P-wave (PP$_i$) propagating through a medium of density, ρ_1, P-wave velocity, V_{p1}, and S-wave velocity, V_{s1}, is incident upon an interface with a medium of density, ρ_2, P-wave velocity, V_{p2}, and S-wave velocity, V_{s2}, at an angle, θ_1. Mode conversions occur resulting in reflected P- and S- waves (PP$_r$ and PS$_r$ respectively) and transmitted (refracted) P- and S- waves (PP$_t$ and PS$_t$ respectively).

Diagram showing the mode conversions that occur when a P-wave reflects off an interface at non-normal incidence

The reflection and transmission coefficients, which govern the amplitude of each reflection, vary with angle of incidence and can be used to obtain information about (among many other things) the fluid content of the rock. Practical use of non-normal incidence phenomena, known as AVO has been facilitated by theoretical work to derive workable approximations to the Zoeppritz equations and by advances in computer processing capacity. AVO studies attempt with some success to predict the fluid content (oil, gas, or water) of potential reservoirs, to lower the risk of drilling unproductive wells and to identify new petroleum reservoirs. The 3-term simplification of the Zoeppritz equations that is most commonly used was developed in 1985 and is known as the "Shuey equation". A further 2-term simplification is known as the "Shuey approximation", is valid for angles of incidence less than 30 degrees (usually the case in seismic surveys) and is given below:

$$R(\theta) = R(0) + G \sin^2 \theta$$

where $R(0)$ = reflection coefficient at zero-offset (normal incidence); G = AVO gradient, describing reflection behaviour at intermediate offsets and (θ) = angle of incidence. This equation reduces to that of normal incidence at (θ) =0.

Interpretation of Reflections

The time it takes for a reflection from a particular boundary to arrive at the geophone is called the *travel time*. If the seismic wave velocity in the rock is known, then the travel time may be used to estimate the depth to the reflector. For a simple vertically traveling wave, the travel time t from the surface to the reflector and back is called the Two-Way Time (TWT) and is given by the formula

$$t = 2\frac{d}{V},$$

where d is the depth of the reflector and V is the wave velocity in the rock.

A series of apparently related reflections on several seismograms is often referred to as a *reflection event*. By correlating reflection events, a seismologist can create an estimated cross-section of the geologic structure that generated the reflections. Interpretation of large surveys is usually performed with programs using high-end three-dimensional computer graphics.

Sources of Noise

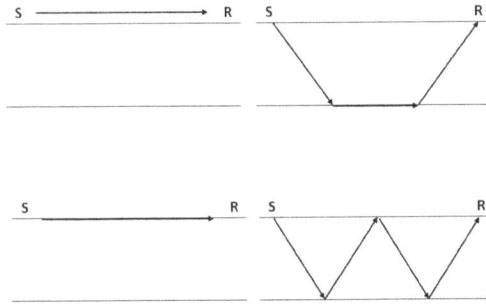

Sources of noise on a seismic record. Top-left: air wave; top-right: head wave; bottom-left: surface wave; bottom-right: multiple.

In addition to reflections off interfaces within the subsuface, there are a number of other seismic responses detected by receivers and are either unwanted or unneeded:

Air Wave

The airwave travels directly from the source to the receiver and is an example of coherent noise. It is easily recognizable because it travels at a speed of 330 m/s, the speed of sound in air.

Ground Roll / Rayleigh wave / Scholte Wave / Surface wave

A Rayleigh wave typically propagates along a free surface of a solid, but the elastic constants and density of air are very low compared to those of rocks so the surface of the Earth is approximately a free surface. Low velocity, low frequency and high amplitude Rayleigh waves are frequently present on a seismic record and can obscure signal, degrading overall data quality. They are known within the industry as 'Ground Roll' and are an example of coherent noise that can be attenuated with a carefully designed seismic survey. The Scholte wave is similar to ground roll but occurs at the sea-floor (fluid/solid interface) and it can possibly obscure and mask deep reflections in marine seismic records. The velocity of these waves varies with wavelength, so they are said to be dispersive and the shape of the wavetrain varies with distance.

Refraction / Head Wave / Conical Wave

A head wave refracts at an interface, travelling along it, within the lower medium and produces oscillatory motion parallel to the interface. This motion causes a disturbance in the upper medium that is detected on the surface. The same phenomenon is utilised in seismic refraction.

Multiple Reflection

An event on the seismic record that has incurred more than one reflection is called a *multiple*. Multiples can be either short-path (peg-leg) or long-path, depending upon whether they interfere with primary reflections or not.

Multiples from the bottom of a body of water (the interface of the base of water and the rock or sediment beneath it) and the air-water interface are common in marine seismic data, and are suppressed by seismic processing.

Cultural Noise

Cultural noise includes noise from weather effects, planes, helicopters, electrical pylons, and ships (in the case of marine surveys), all of which can be detected by the receivers.

Applications

Reflection seismology is used extensively in a number of fields and its applications can be categorised into three groups, each defined by their depth of investigation:

- Near-surface applications – an application that aims to understand geology at depths of up to approximately 1 km, typically used for engineering and environmental surveys, as well as coal and mineral exploration. A more recently developed application for seismic reflection is for geothermal energy surveys, although the depth of investigation can be up to 2 km deep in this case.

- Hydrocarbon exploration - used by the hydrocarbon industry to provide a high resolution map of acoustic impedance contrasts at depths of up to 10 km within the subsurface. This can be combined with seismic attribute analysis and other exploration geophysics tools and used to help geologists build a geological model of the area of interest.

- Crustal studies – investigation into the structure and origin of the Earth's crust, through to the Moho discontinuity and beyond, at depths of up to 100 km.

A method similar to reflection seismology which uses electromagnetic instead of elastic waves, and has a smaller depth of penetration, is known as Ground-penetrating radar or GPR.

Hydrocarbon Exploration

Reflection seismology, more commonly referred to as "seismic reflection" or abbreviated to "seismic" within the hydrocarbon industry, is used by petroleum geologists and geophysicists to map and interpret potential petroleum reservoirs. The size and scale of seismic surveys has increased alongside the significant concurrent increases in computer power during the last 25 years. This has led the seismic industry from laboriously – and therefore rarely – acquiring small 3D surveys in the 1980s to now routinely acquiring large-scale high resolution 3D surveys. The goals and basic principles have remained the same, but the methods have slightly changed over the years.

Seismic testing in 1940.

The primary environments for seismic exploration are land, the transition zone and marine:

Land - The land environment covers almost every type of terrain that exists on Earth, each bringing its own logistical problems. Examples of this environment are jungle, desert, arctic tundra, forest, urban settings, mountain regions and savannah.

Transition Zone (TZ) - The transition zone is considered to be the area where the land meets the sea, presenting unique challenges because the water is too shallow for large seismic vessels but too deep for the use of traditional methods of acquisition on land. Examples of this environment are river deltas, swamps and marshes, coral reefs, beach tidal areas and the surf zone. Transition zone seismic crews will often work on land, in the transition zone and in the shallow water marine environment on a single project in order to obtain a complete map of the subsurface.

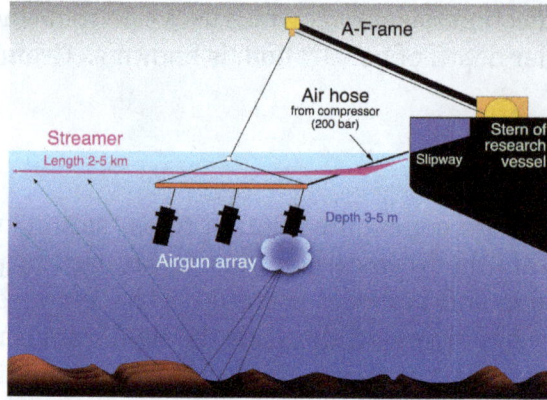

Diagram of equipment used for marine seismic surveys

Marine - The marine zone is either in shallow water areas (water depths of less than 30 to 40 metres would normally be considered shallow water areas for 3D marine seismic operations) or in the deep water areas normally associated with the seas and oceans (such as the Gulf of Mexico).

Seismic surveys are typically designed by National oil companies and International oil companies who hire service companies such as Breckenridge Exploration Co., CGG, Petroleum Geo-Services and WesternGeco to acquire them. Another company is then hired to process the data, although this can often be the same company that acquired the survey. Finally the finished seismic volume is delivered to the oil company so that it can be geologically interpreted.

Land Survey Acquisition

Land seismic surveys tend to be large entities, requiring hundreds of tons of equipment and employing anywhere from a few hundred to a few thousand people, deployed over vast areas for many months. There are a number of options available for a controlled seismic source in a land survey and particularly common choices are Vibroseis and dynamite. Vibroseis is a non-impulsive source that is cheap and efficient but requires flat ground to operate on, making its use more difficult in undeveloped areas. The method comprises one or more heavy, all-terrain vehicles lowering a steel plate onto the ground, which is then vibrated with a specific frequency distribution and amplitude. It produces a low energy density, allowing it to be used in cities and other built-up areas where dynamite would cause significant damage, though the large weight attached to a Vibroseis truck can cause its own environmental damage. Dynamite is an impulsive source that is regarded as the ideal geophysical source due to it producing an almost perfect impulse function but it has obvious environmental drawbacks. For a long time, it was the only seismic source available until weight dropping was introduced around 1954, allowing geophysicists to make a trade-off between image quality and environmental damage. Compared to Vibroseis, dynamite is also operationally inefficient because each source point needs to be drilled and the dynamite placed in the hole.

A land seismic survey requires substantial logistical support. In addition to the day-to-day seismic operation itself, there must also be support for the main camp (for catering, waste management and laundry etc.), smaller camps (for example where the distance is too far to drive back to the main camp with vibrator trucks), vehicle and equipment maintenance, medical personnel and security.

Desert land seismic camp

Receiver line on a desert land crew with recorder truck

Unlike in marine seismic surveys, land geometries are not limited to narrow paths of acquisition, meaning that a wide range of offsets and azimuths is usually acquired and the largest challenge is increasing the rate of acquisition. The rate of production is obviously controlled by how fast the source (Vibroseis in this case) can be fired and then move on to the next source location. Attempts have been made to use multiple seismic sources at the same time in order to increase survey efficiency and a successful example of this technique is Independent Simultaneous Sweeping (ISS).

Marine Survey Acquisition (Streamer)

Traditional marine seismic surveys are conducted using specially-equipped vessels that tow one or more cables containing a series of hydrophones at constant intervals. The cables are known as *streamers*, with 2D surveys using only 1 streamer and 3D surveys employing up to 12 or more (though 6 or 8 is more common). The streamers are deployed just beneath the surface of the water and are at a set distance away from the vessel. The seismic source, usually an airgun or an array of airguns but other sources are available, is also deployed beneath the water surface and is located between the vessel and the first receiver. Two identical sources are often used to achieve a faster rate of shooting. Marine seismic surveys generate a significant quantity of data, each streamer can be up to 6 or even 8 km long, containing hundreds of channels and the seismic source is typically fired every 15 or 20 seconds.

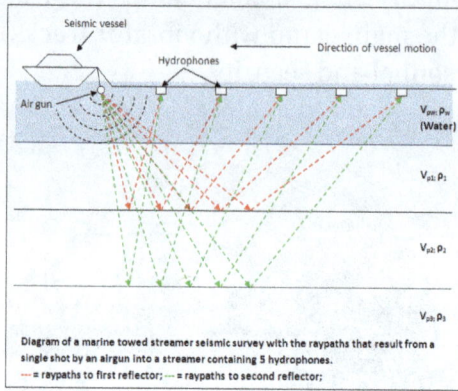

Diagram of a marine towed streamer seismic survey with the raypaths that result from a single shot by an airgun into a streamer containing 5 hydrophones.
--- = raypaths to first reflector; --- = raypaths to second reflector;

Marine seismic survey using a towed streamer

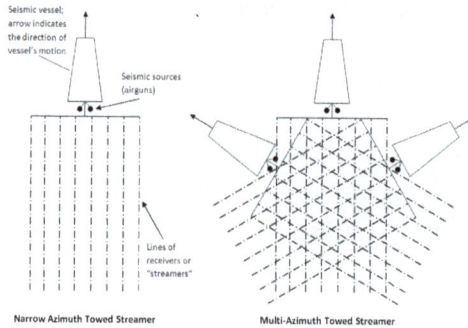

Narrow Azimuth Towed Streamer   Multi-Azimuth Towed Streamer

Plan view of NATS and MAZ surveys

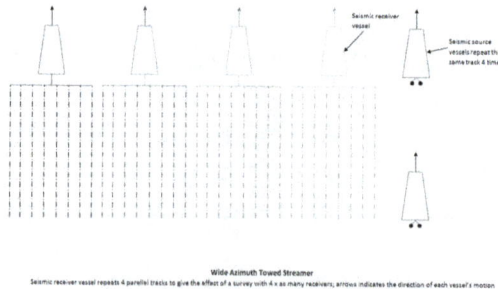

Wide-Azimuth Towed Streamer
Seismic receiver vessel repeats 4 parallel tracks to give the effect of a survey with 4 x as many receivers; arrows indicates the direction of each vessel's motion

Plan view of a WATS/WAZ survey

A seismic vessel with 2 sources and towing a single streamer is known as a *Narrow-Azimuth Towed Streamer* (or NAZ or NATS). By the early 2000s, it was accepted that this type of acquisition was useful for initial exploration but inadequate for development and production, in which wells had to be accurately positioned. This led to the development of the *Multi-Azimuth Towed Streamer* (MAZ) which tried to break the limitations of the linear acquisition pattern of a NATS survey by acquiring a combination of NATS surveys at different azimuths. This successfully delivered increased illumination of the subsurface and a better signal to noise ratio.

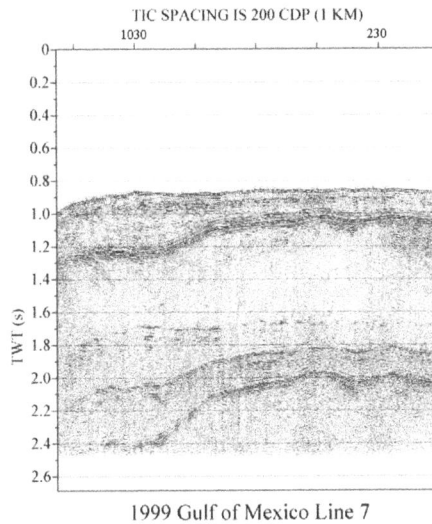

1999 Gulf of Mexico Line 7

Seismic data collected by the USGS in the Gulf of Mexico

The seismic properties of salt poses an additional problem for marine seismic surveys, it attenuates seismic waves and its structure contains overhangs that are difficult to image. This led to another variation on the NATS survey type, the *wide-azimuth towed streamer* (or WAZ or WATS) and was first tested on the Mad Dog field in 2004. This type of survey involved 1 vessel solely towing a set of 8 streamers and 2 separate vessels towing seismic sources that were located at the start and end of the last receiver line. This configuration was "tiled" 4 times, with the receiver vessel moving further away from the source vessels each time and eventually creating the effect of a survey with 4 times the number of streamers. The end result was a seismic dataset with a larger range of wider azimuths, delivering a breakthrough in seismic imaging. These are now the three common types of marine towed streamer seismic surveys.

Marine Survey Acquisition (Ocean-bottom & 4D)

Marine survey acquisition is not just limited to seismic vessels; it is also possible to lay cables of geophones and hydrophones on the sea bed in a similar way to how cables are used in a land seismic survey, and use a separate source vessel. This method was originally developed out of operational necessity in order to enable seismic surveys to be conducted in areas with obstructions, such as production platforms, without having the compromise the resultant image quality. Ocean bottom cables (OBC) are also extensively used in other areas that a seismic vessel cannot be used, for example in shallow marine (water depth <300m) and transition zone environments, and can be deployed by ROVs in deep water when repeatability is valued. Conventional OBC surveys use dual-component receivers, combining a pressure sensor (hydrophone) and a vertical particle velocity sensor (vertical geophone), but more recent developments have expanded the method to use four-component sensors i.e. a hydrophone and three orthogonal geophones. Four-component sensors have the advantage of being able to also record shear waves, which do not travel through water but can still contain valuable information.

In addition to the operational advantages, OBC also has geophysical advantages over a conventional NATS survey that arise from the increased fold and wider range of azimuths associated with the survey geometry. However, much like a land survey, the wider azimuths and increased fold come at a cost and the ability for large-scale OBC surveys is severely limited.

4D surveys are 3D seismic surveys repeated over a period of time in order to observe reservoir depletion during production and identify areas where there are barriers to flow that may not be easily detectable in conventional seismic. They are conventionally conducted using ocean-bottom cables because the cables can be accurately placed in their previous location after being removed. A number of 4D surveys have also been set up over fields in which ocean bottom cables have been purchased and permanently deployed. This method can be known as Life of Field Seismic (LoF).

In 2005, Ocean Bottom Nodes/Seismic (OBN / OBS) - an extension of the OBC method that uses battery-powered cableless receivers placed in deep water - was first trialled over the Atlantis Oil Field in a partnership between BP and Fairfield Industries. The placement of these nodes can be more flexible than the cables in OBC and they are easier to store and deploy due to their smaller size and lower weight. The world's first 4D survey using nodes was acquired over the Atlantis field in 2009, with the nodes being placed by a ROV in a water depth of 1300-2200m to within 30m of where they were previously placed in 2005.

Seismic Data Processing

There are three main processes in seismic data processing: deconvolution, common-midpoint (CMP) stacking and migration.

Deconvolution is a process that tries to extract the reflectivity series of the Earth, under the assumption that a seismic trace is just the reflectivity series of the Earth convolved with distorting filters. This process improves temporal resolution by collapsing the seismic wavelet, but it is nonunique unless further information is available such as well logs, or further assumptions are made. *Deconvolution* operations can be cascaded, with each individual deconvolution designed to remove a particular type of distortion.

CMP stacking is a robust process that uses the fact that a particular location in the subsurface will have been sampled numerous times and at different offsets. This allows a geophysicist to construct a group of traces with a range of offsets that all sample the same subsurface location, known as a *Common Midpoint Gather*. The average amplitude is then calculated along a time sample, resulting in significantly lowering the random noise but also losing all valuable information about the relationship between seismic amplitude and offset. Less significant processes that are applied shortly before the *CMP stack* are *Normal moveout correction* and *statics correction*. Unlike marine seismic data, land seismic data has to be corrected for the elevation differences between the shot and receiver locations. This correction is in the form of a vertical time shift to a flat datum and is known as a *statics correction*, but will need further correcting later in the processing sequence because the velocity of the near-surface is not accurately known. This further correction is known as a *residual statics correction*.

Seismic migration is the process by which seismic events are geometrically re-located in either space or time to the location the event occurred in the subsurface rather than the location that it was recorded at the surface, thereby creating a more accurate image of the subsurface.

Seismic Interpretation

The goal of seismic interpretation is to obtain a coherent geological story from the map of processed seismic reflections. At its most simple level, seismic interpretation involves tracing and correlating along continuous reflectors throughout the 2D or 3D dataset and using these as the basis for the geological interpretation. The aim of this is to produce structural maps that reflect the spatial variation in depth of certain geological layers. Using these maps hydrocarbon traps can be identified and models of the subsurface can be created that allow volume calculations to be made. However, a seismic dataset rarely gives a picture clear enough to do this. This is mainly because of the vertical and horizontal seismic resolution but often noise and processing difficulties also result in a lower quality picture. Due to this, there is always a degree of uncertainty in a seismic interpretation and a particular dataset could have more than one solution that fits the data. In such a case, more data will be needed to constrain the solution, for example in the form of further seismic acquisition, borehole logging or gravity and magnetic survey data. Similarly to the mentality of a seismic processor, a seismic interpreter is generally encouraged to be optimistic in order encourage further work rather than the abandonment of the survey area. Seismic interpretation is completed by both geologists and geophysicists, with most seismic interpreters having an understanding of both fields.

In hydrocarbon exploration, the features that the interpreter is particularly trying to delineate are the parts that make up a petroleum reservoir - the source rock, the reservoir rock, the seal and trap.

Seismic Attribute Analysis

Seismic attribute analysis involves extracting or deriving a quantity from seismic data that can be analysed in order to enhance information that might be more subtle in a traditional seismic image, leading to a better geological or geophysical interpretation of the data. Examples of attributes that can be analysed include mean amplitude, which can lead to the delineation of bright spots and dim spots, coherency and amplitude versus offset. Attributes that can show the presence of hydrocarbons are called direct hydrocarbon indicators.

Crustal Studies

The use of reflection seismology in studies of tectonics and the Earth's crust was pioneered in the 1970s by groups such as the Consortium for Continental Reflection Profiling (COCORP), who inspired deep seismic exploration in other countries such as BIRPS in Great Britain and ECORS in France. The British Institutions Reflection Profiling Syndicate (BIRPS) was started up as a result of oil hydrocarbon exploration in the North Sea. It became clear that there was a lack of understanding of the tectonic processes that had formed the geological structures and sedimentary basins which were being explored. The effort produced some significant results and showed that it is possible to profile features such as thrust faults that penetrate through the crust to the upper mantle with marine seismic surveys.

Environmental Impact

As with all human activities, seismic reflection surveys may have some impact on the Earth's natural environment and both the hydrocarbon industry and environmental groups partake in research to investigate these effects.

Land

On land, conducting a seismic survey may require the building of roads, for transporting equipment and personnel, and vegetation may need to be cleared for the deployment of equipment. If the survey is in a relatively undeveloped area, significant habitat disturbance may occur and many governments require seismic companies to follow strict rules regarding destruction of the environment; for example, the use of dynamite as a seismic source may be disallowed. Seismic processing techniques allow for seismic lines to deviate around natural obstacles, or use pre-existing non-straight tracks and trails. With careful planning, this can greatly reduce the environmental impact of a land seismic survey. The more recent use of inertial navigation instruments for land survey instead of theodolites decreased the impact of seismic by allowing the winding of survey lines between trees.

Marine

The main environmental concern for marine seismic surveys is the potential for noise associated with the high-energy seismic source to disturb or injure animal life, especially cetaceans such as whales, porpoises, and dolphins, as these mammals use sound as their primary method of communication with one another. High-level and long-duration sound can cause physical damage, such as hearing loss, whereas lower-level noise can cause temporary threshold shifts in hearing, obscuring sounds that are vital to marine life, or behavioural disturbance.

A study has shown that migrating humpback whales will leave a minimum 3 km gap between themselves and an operating seismic vessel, with resting humpback whale pods with cows exhibiting increased sensitivity and leaving an increased gap of 7–12 km. Conversely, the study found that male humpback whales were attracted to a single operating airgun as they were believed to have confused the low-frequency sound with that of whale breaching behaviour. In addition to whales, sea turtles, fish and squid all showed alarm and avoidance behaviour in the presence of an approaching seismic source. It is difficult to compare reports on the effects of seismic survey noise on marine life because methods and units are often inadequately documented.

The gray whale will avoid its regular migratory and feeding grounds by >30 km in areas of seismic testing. Similarly the breathing of gray whales was shown to be more rapid, indicating discomfort and panic in the whale. It is circumstantial evidence such as this that has led researchers to believe that avoidance and panic might be responsible for increased whale beachings although research is ongoing into these questions.

Offering another point of view, a joint paper from the International Association of Geophysical Contractors (IAGC) and the International Association of Oil and Gas Producers (OGP) argue that the noise created by marine seismic surveys is comparable to natural sources of seismic noise, stating:

"The sound produced during seismic surveys is comparable in magnitude to many naturally occurring and other man-made sound sources. Furthermore, the specific characteristics of seismic sounds and the operational procedures employed during seismic surveys are such that the resulting risks to marine mammals are expected to be exceptionally low. In fact, three decades of world-wide seismic surveying activity and a variety of research projects have shown no evidence which would suggest that sound from E&P seismic activities has resulted in any physical or auditory injury to any marine mammal species."

History

Reflections and refractions of seismic waves at geologic interfaces within the Earth were first observed on recordings of earthquake-generated seismic waves. The basic model of the Earth's deep interior is based on observations of earthquake-generated seismic waves transmitted through the Earth's interior (e.g., Mohorovičić, 1910). The use of human-generated seismic waves to map in detail the geology of the upper few kilometers of the Earth's crust followed shortly thereafter and has developed mainly due to commercial enterprise, particularly the petroleum industry.

Seismic reflection exploration grew out of the seismic refraction exploration method, which was used to find oil associated with salt domes. Ludger Mintrop, a German mine surveyor, devised a mechanical seismograph in 1914 that he successfully used to detect salt domes in Germany. He applied for a German patent in 1919 that was issued in

1926. In 1921 he founded the company Seismos, which was hired to conduct seismic exploration in Texas and Mexico, resulting in the first commercial discovery of oil using the refraction seismic method in 1924. The 1924 discovery of the Orchard salt dome in Texas led to a boom in seismic refraction exploration along the Gulf Coast, but by 1930 the method had led to the discovery most of the shallow Gulf Coast salt domes, and the refraction seismic method faded.

The Canadian inventor Reginald Fessenden was the first to conceive of using reflected seismic waves to infer geology. His work was initially on the propagation of acoustic waves in water, motivated by the sinking of the Titanic by an iceberg in 1912. He also worked on methods of detecting submarines during World War I. He applied for the first patent on a seismic exploration method in 1914, which was issued in 1917. Due to the war, he was unable to follow up on the idea. John Clarence Karcher discovered seismic reflections independently while working for the United States Bureau of Standards (now the National Institute of Standards and Technology) on methods of sound ranging to detect artillery. In discussion with colleagues, the idea developed that these reflections could aid in exploration for petroleum. With several others, many affiliated with the University of Oklahoma, Karcher helped to form the Geological Engineering Company, incorporated in Oklahoma in April, 1920. The first field tests were conducted near Oklahoma City, Oklahoma in 1921.

Early reflection seismology was viewed with skepticism by many in the oil industry. An early advocate of the method commented:

> "As one who personally tried to introduce the method into general consulting practice, the senior writer can definitely recall many times when reflections were not even considered on a par with the divining rod, for at least that device had a background of tradition."

The Geological Engineering Company folded due to a drop in the price of oil. In 1925, oil prices had rebounded, and Karcher helped to form Geophysical Research Corporation (GRC) as part of the oil company Amerada. In 1930, Karcher left GRC and helped to found Geophysical Service Incorporated (GSI). GSI was one of the most successful seismic contracting companies for over 50 years and was the parent of an even more successful company, Texas Instruments. Early GSI employee Henry Salvatori left that company in 1933 to found another major seismic contractor, Western Geophysical. Many other companies using reflection seismology in hydrocarbon exploration, hydrology, engineering studies, and other applications have been formed since the method was first invented. Major service companies today include Dawson Geophysical, WesternGeco, CGG, TGS, ION Geophysical, Petroleum Geo-Services and Polarcus. Most major oil companies also have actively conducted research into seismic methods as well as collected and processed seismic data using their own personnel and technology. Reflection seismology has also found applications in non-commercial research by academic and government scientists around the world.

Seismic Source

A seismic source is a device that generates controlled seismic energy used to perform both reflection and refraction seismic surveys. A seismic source can be simple, such as dynamite, or it can use more sophisticated technology, such as a specialized air gun. Seismic sources can provide single pulses or continuous sweeps of energy, generating seismic waves, which travel through a medium such as water or layers of rocks. Some of the waves then reflect and refract and are recorded by receivers, such as geophones or hydrophones.

An air gun seismic source (30 litre)

Seismic sources may be used to investigate shallow subsoil structure, for engineering site characterization, or to study deeper structures, either in the search for petroleum and mineral deposits, or to map subsurface faults or for other scientific investigations. The returning signals from the sources are detected by seismic sensors (geophones or hydrophones) in known locations relative to the position of the source. The recorded signals are then subjected to specialist processing and interpretation to yield comprehensible information about the subsurface.

Source Model

A seismic source signal has the following characteristics:

1. Generates an impulse signal

2. Band-limited

3. The generated waves are time-varying

The generalized equation that shows all above properties is:

$$s(t) = \beta e^{-\alpha t^2} \sin(2\pi f_{max} t)$$

where f_{max} is the maximum frequency component of the generated waveform.

Types of Sources

Explosives

Explosives, such as dynamite, can be used as crude but effective sources of seismic energy. For instance, hexanitrostilbene was the main explosive fill in the *thumper* mortar round canisters used as part of the Apollo Lunar Active Seismic Experiments. Generally, the explosive charges are placed between 6 and 76 metres (20 and 250 ft) below ground, in a hole that is drilled with dedicated drilling equipment for this purpose. This type of seismic drilling is often referred to as "Shot Hole Drilling". A common drill rig used for "Shot Hole Drilling" is the ARDCO C-1000 drill mounted on an ARDCO K 4X4 buggy. These drill rigs often use water or air to assist the drilling.

Air Gun

An air gun is used for marine reflection and refraction surveys. It consists of one or more pneumatic chambers that are pressurized with compressed air at pressures from 14 to 21 MPa (2,000 to 3,000 psi). Air guns are submerged below the water surface, and towed behind a ship. When an air gun is fired, a solenoid is triggered, which releases air into a fire chamber which in turn causes a piston to move, thereby allowing the air to escape the main chamber and producing a pulse of acoustic energy. Air gun arrays may consist of up to 48 individual air guns with different size chambers, fired in concert, the aim being to create the optimum initial shock wave followed by the minimum reverberation of the air bubble(s).

Air guns are made from the highest grades of corrosion resistant stainless steel. Large chambers (i.e., greater than 1.15 L or 70 cu in) tend to give low frequency signals, and the small chambers (less than 70 cubic inches) give higher frequency signals.

Plasma Sound Source

A plasma sound source (PSS), otherwise called a spark gap sound source, or simply a sparker, is a means of making a very low frequency sonar pulse underwater. For each firing, electric charge is stored in a large high-voltage bank of capacitors, and then released in an arc across electrodes in the water. The underwater spark discharge produces a high-pressure plasma and vapor bubble, which expands and collapses, making a loud sound. Most of the sound produced is between 20 and 200 Hz, useful for both seismic and sonar applications.

There are also plans to use PSS as a non-lethal weapon against submerged divers.

Thumper Truck

Thumper trucks, Noble Energy, northern Nevada 2012.

In 1953, the weight dropping Thumper technique was introduced as an alternative to dynamite sources.

A thumper truck (or weight-drop) truck is a vehicle-mounted ground impact system which can be used to provide a seismic source. A heavy weight is raised by a hoist at the back of the truck and dropped, generally about three meters, to impact (or "thump") the ground. To augment the signal, the weight may be dropped more than once at the same spot, the signal may also be increased by thumping at several nearby places in an array whose dimensions may be chosen to enhance the seismic signal by spatial filtering.

More advanced Thumpers use a technology called "Accelerated Weight Drop" (AWD), where a high pressure gas (min 6.9 MPa (1,000 psi)) is used to accelerate a heavy weight Hammer (5,000 kg) to hit a base plate coupled to the ground from a distance of 2 to 3 m. Several thumps are stacked to enhance signal to noise ratio. AWD allows both more energy and more control of the source than gravitational weight-drop, providing better depth penetration, control of signal frequency content.

Thumping may be less damaging to the environment than firing explosives in shot-holes, though a heavily thumped seismic line with transverse ridges every few meters might create long-lasting disturbance of the soil. An advantage of the thumper (later shared with Vibroseis), especially in politically unstable areas, is that no explosives are required.

Electromagnetic Pulse Energy Source (Non-Explosive)

EMP sources based on the electrodynamic and electromagnetic principles.

Seismic Vibrator

A Seismic vibrator propagates energy signals into the Earth over an extended period of time as opposed to the near instantaneous energy provided by impulsive sources.

The data recorded in this way must be *correlated* to convert the extended source signal into an impulse. The source signal using this method was originally generated by a servo-controlled hydraulic vibrator or *shaker unit* mounted on a mobile base unit, but electro-mechanical versions have also been developed.

The "Vibroseis" exploration technique was developed by the Continental Oil Company (Conoco) during the 1950s and was a trademark until the company's patent lapsed.

Boomer Sources

Boomer sound sources are used for shallow water seismic surveys, mostly for engineering survey applications. Boomers are towed in a floating sled behind a survey vessel. Similar to the plasma source, a boomer source stores energy in capacitors, but it discharges through a flat spiral coil instead of generating a spark. A copper plate adjacent to the coil flexes away from the coil as the capacitors are discharged. This flexing is transmitted into the water as the seismic pulse.

Originally the storage capacitors were placed in a steel container (the **bang box**) on the survey vessel. The high voltages used, typically 3,000 V, required heavy cables and strong safety containers. Recently, low voltage boomers have become available. These use capacitors on the towed sled, allowing efficient energy recovery, lower voltage power supplies and lighter cables. The low voltage systems are generally easier to deploy and have fewer safety concerns.

Noise Sources

Correlation-based processing techniques also enable seismologists to image the interior of the Earth at multiple scales using natural (e.g., the oceanic microseism) or artificial (e.g., urban) background noise as a seismic source. For example, under ideal conditions of uniform seismic illumination, the correlation of the noise signals between two seismographs provides an estimate of the bidirectional seismic impulse response.

Seismic Migration

Seismic migration is the process by which seismic events are geometrically re-located in either space or time to the location the event occurred in the subsurface rather than the location that it was recorded at the surface, thereby creating a more accurate image of the subsurface. This process is necessary to overcome the limitations of geophysical methods imposed by areas of complex geology, such as: faults, salt bodies, folding, etc.

Migration moves dipping reflectors to their true subsurface positions and collapses diffractions, resulting in a migrated image that typically has an increased spatial resolution and resolves areas of complex geology much better than non-migrated images. A form of migration is one of the standard data processing techniques for reflection-based geophysical methods (seismic reflection and ground-penetrating radar)

The need for migration has been understood since the beginnings of seismic exploration and the very first seismic reflection data from 1921 were migrated. Computational migration algorithms have been around for many years but they have only entered wide usage in the past 20 years because they are extremely resource-intensive. Migration can lead to a dramatic uplift in image quality so algorithms are the subject of intense research, both within the geophysical industry as well as academic circles.

Rationale

Seismic waves are elastic waves that propagate through the Earth with a finite velocity, governed by the acoustic properties of the rock in which they are travelling. At an interface between two rock types, with different acoustic impedances, the seismic energy is either refracted, reflected back towards the surface or attenuated by the medium. The reflected energy arrives at the surface and is recorded by geophones that are placed at a known distance away from the source of the waves. When a geophysicist views the recorded energy from the geophone, they know both the travel time and the distance between the source and the receiver, but not the distance down to the reflector. In the simplest geological setting, with a single horizontal reflector, a constant velocity and a source and receiver at the same location (referred to as zero-offset, where offset is the distance between the source and receiver), the geophysicist can determine the location of the reflection event by using the relationship:

$$d = \frac{vt}{2},$$

where d is the distance, v is the seismic velocity (or rate of travel) and t is the measured time from the source to the receiver.

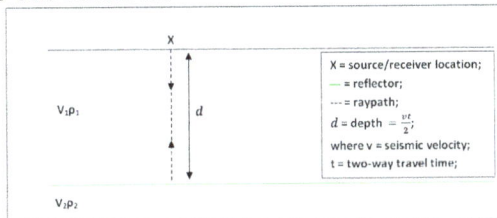

Diagram showing the raypath for a zero-offset reflection from a horizontal reflector.

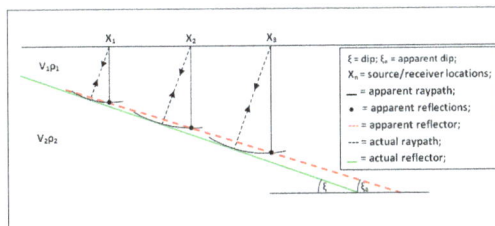

Diagram showing the raypath for a zero-offset reflection from a dipping reflector and the resultant apparent dip.

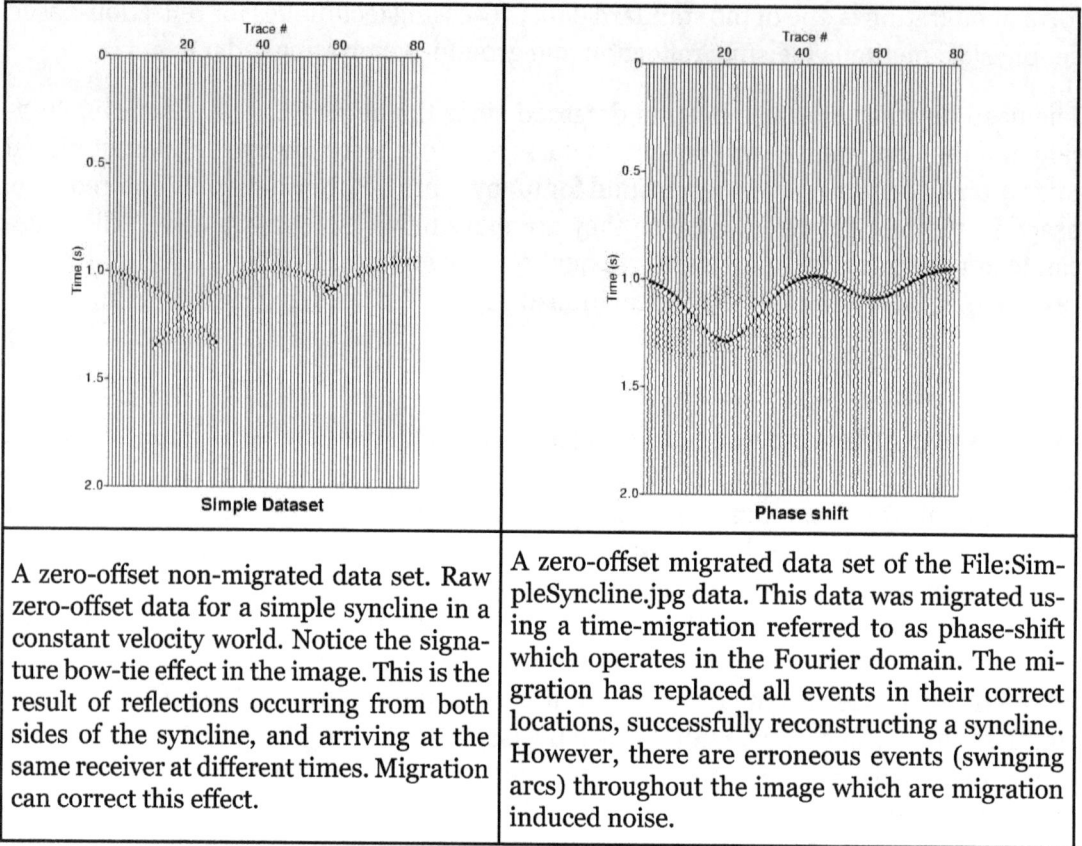

A zero-offset non-migrated data set. Raw zero-offset data for a simple syncline in a constant velocity world. Notice the signature bow-tie effect in the image. This is the result of reflections occurring from both sides of the syncline, and arriving at the same receiver at different times. Migration can correct this effect.	A zero-offset migrated data set of the File:SimpleSyncline.jpg data. This data was migrated using a time-migration referred to as phase-shift which operates in the Fourier domain. The migration has replaced all events in their correct locations, successfully reconstructing a syncline. However, there are erroneous events (swinging arcs) throughout the image which are migration induced noise.

In this case, the distance is halved because it can be assumed that it only took one-half of the total travel time to reach the reflector from the source, then the other half to return to the receiver.

The result gives us a single scalar value, which actually represents a half-sphere of distances, from the source/receiver, which the reflection could have originated from. It is a half-sphere, and not a full sphere, because we can ignore all possibilities that occur above the surface as unreasonable. In the simple case of a horizontal reflector, it can be assumed that the reflection is located vertically below the source/receiver point.

The situation is more complex in the case of a dipping reflector, as the first reflection originates from further up the direction of dip and therefore the travel-time plot will show a reduced dip that is defined the "migrator's equation" :

$$\tan \xi_a = \sin \xi,$$

where ξ_a is the *apparent dip* and ξ is the *true dip*.

Zero-offset data is important to a geophysicist because the migration operation is much simpler, and can be represented by spherical surfaces. When data is acquired at non-zero offsets, the sphere becomes an ellipsoid and is much more complex to represent (both geometrically, as well as computationally).

Use

For a geophysicist, complex geology is defined as anywhere there is an abrupt or sharp contrast in lateral and/or vertical velocity (e.g. a sudden change in rock type or lithology which causes a sharp change in seismic wave velocity).

Some examples of what a geophysicist considers complex geology are: faulting, folding, (some) fracturing, salt bodies, and unconformities. In these situations a form of migration is used called pre-stack migration (PreSM), in which all traces are migrated before being moved to zero-offset. Consequently, much more information is used, which results in a much better image, along with the fact that PreSM honours velocity changes more accurately than post-stack migration.

Types of Migration

Depending on budget, time restrictions and the subsurface geology, geophysicists can employ 1 of 2 fundamental types of migration algorithms, defined by the domain in which they are applied: time migration and depth migration.

Time Migration

Time Migration is applied to seismic data in time coordinates.This type of migration makes the assumption of only mild lateral velocity variations and this breaks down in the presence of most interesting and complex subsurface structures, particularly salt. Some popularly used time migration algorithms are: Stolt migration, Gazdag and Finite-difference migration.

Depth Migration

Depth Migration is applied to seismic data in depth (regular Cartesian) coordinates, which must be calculated from seismic data in time coordinates. This method does therefore require a velocity model, making it resource-intensive because building a seismic velocity model is a long and iterative process. The significant advantage to this migration method is that it can be successfully used in areas with lateral velocity variations, which tend to be the areas that are most interesting to petroleum geologists. Some of the popularly used depth migration algorithms are Kirchhoff depth migration, Reverse Time Migration (RTM), Gaussian Beam Migration and Wave-equation migration.

Resolution

The goal of migration is to ultimately increase spatial resolution and one of the basic assumptions made about the seismic data is that it only shows primary reflections and all noise has been removed. In order to ensure maximum resolution (and therefore maximum uplift in image quality) the data should be sufficiently pre-processed before migration. Noise that may be easy to distinguish pre-migration could be smeared across the entire aperture length during migration, reducing image sharpness and clarity.

A further basic consideration is whether to use 2D or 3D migration. If the seismic data has an element of cross-dip (a layer that dips perpendicular to the line of acquisition) then the primary reflection will originate from out-of-plane and 2D migration cannot put the energy back to its origin. In this case, 3D migration is needed to attain the best possible image.

Modern seismic processing computers are more capable of performing 3D migration, so the question of whether to allocate resources to performing 3D migration is less of a concern.

Graphical Migration

The simplest form of migration is that of graphical migration. Graphical migration assumes a constant velocity world and zero-offset data, in which a geophysicist draws spheres or circles from the receiver to the event location for all events. The intersection of the circles then form the reflector's "true" location in time or space. An example of such can be seen in the diagram.

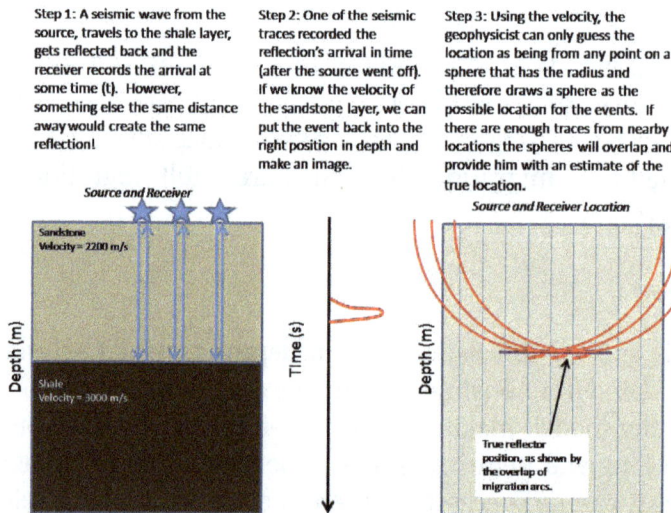

Step 1: A seismic wave from the source, travels to the shale layer, gets reflected back and the receiver records the arrival at some time (t). However, something else the same distance away would create the same reflection!

Step 2: One of the seismic traces recorded the reflection's arrival in time (after the source went off). If we know the velocity of the sandstone layer, we can put the event back into the right position in depth and make an image.

Step 3: Using the velocity, the geophysicist can only guess the location as being from any point on a sphere that has the radius and therefore draws a sphere as the possible location for the events. If there are enough traces from nearby locations the spheres will overlap and provide him with an estimate of the true location.

An example of simple graphical migration. Until the advent of modern computers in the 1960s and 1970s this was a method used by geophysicists to primitively 'migrate' their data. This method is obsolete with the advent of digital processors, but is useful for understanding the basic principle behind migration.

Technical Details

Migration of seismic data is the correction of the flat-geological-layer assumption by a numerical, grid-based spatial convolution of the seismic data to account for dipping events (where geological layers are not flat). There are many approaches, such as the popular Kirchhoff migration, but it is generally accepted that processing large spatial sections (apertures) of the data at a time introduces fewer errors, and that depth migration is far superior to time migration with large dips and with complex salt bodies.

Basically, it repositions/moves the energy (seismic data) from the recorded locations to the locations with the correct common midpoint (CMP). While the seismic data is received at the proper locations originally (according to the laws of nature), these locations do not correspond with the assumed CMP for that location. Though stacking the data without the migration corrections yields a somewhat inaccurate picture of the subsurface, migration is preferred for better most imaging recorder to drill and maintain oilfields. This process is a central step in the creation of an image of the subsurface from active source seismic data collected at the surface, seabed, boreholes, etc., and therefore is used on industrial scales by oil and gas companies and their service providers on digital computers.

Explained in another way, this process attempts to account for wave dispersion from dipping reflectors and also for the spatial and directional seismic wave speed (heterogeneity) variations, which cause wavefields (modelled by ray paths) to bend, wave fronts to cross (caustics), and waves to be recorded at positions different from those that would be expected under straight ray or other simplifying assumptions. Finally, this process often attempts to also preserve and extract the formation interface reflectivity information imbedded in the seismic data amplitudes, so that they can be used to reconstruct the elastic properties of the geological formations (amplitude preservation, seismic inversion). There are a variety of migration algorithms, which can be classified by their output domain into the broad categories of Time Migration or Depth Migration, and Pre-Stack Migration or Post-Stack migration (orthogonal) techniques. Depth migration begins with time data converted to depth data by a spatial geological velocity profile. Post-Stack migration begins with seismic data which has already been stacked, and thus already lost valuable velocity analysis information.

Seismic Attribute

In reflection seismology, a seismic attribute is a quantity extracted or derived from seismic data that can be analysed in order to enhance information that might be more subtle in a traditional seismic image, leading to a better geological or geophysical interpretation of the data. Examples of seismic attributes can include measured time, amplitude, frequency and attenuation, in addition to combinations of these. Most seismic attributes are post-stack, but those that use CMP gathers, such as amplitude versus offset (AVO), must be analysed pre-stack. They can be measured along a single seismic trace or across multiple traces within a defined window.

The first attributes developed were related to the 1D complex seismic trace and included: envelope amplitude, instantaneous phase, instantaneous frequency, and apparent polarity. Acoustic impedance obtained from seismic inversion can also be considered an attribute and was among the first developed.

Other attributes commonly used include: coherence, azimuth, dip, instantaneous amplitude, response amplitude, response phase, instantaneous bandwidth, AVO, and spectral decomposition.

A seismic attribute that can indicate the presence or absence of hydrocarbons is known as a direct hydrocarbon indicator.

Amplitude Attributes

Amplitude attributes use the seismic signal amplitude as the basis for their computation.

Mean Amplitude

A post-stack attribute that computes the arithmetic mean of the amplitudes of a trace within a specified window. This can be used to observe the trace bias which could indicate the presence of a bright spot.

Average Energy

A post-stack attribute that computes the sum of the squared amplitudes divided by the number of samples within the specified window used. This provides a measure of reflectivity and allows one to map direct hydrocarbon indicators within a zone of interest.

RMS (Root Mean Square) Amplitude

A post-stack attribute that computes the square root of the sum of squared amplitudes divided by the number of samples within the specified window used. With this root mean square amplitude, one can measure reflectivity in order to map direct hydrocarbon indicators in a zone of interest. However, RMS is sensitive to noise as it squares every value within the window.

Maximum Magnitude

A post-stack attribute that computes the maximum value of the absolute value of the amplitudes within a window. This can be used to map the strongest direct hydrocarbon indicator within a zone of interest.

AVO Attributes

AVO (amplitude versus offset) attributes are pre-stack attributes that have as the basis for their computation, the variation in amplitude of a seismic reflection with varying offset. These attributes include: AVO intercept, AVO gradient, intercept multiplied by gradient, far minus near, fluid factor, etc.

Anelastic Attenuation Factor

The anelastic attenuation factor (or Q) is a seismic attribute that can be determined from seismic reflection data for both reservoir characterisation and advanced seismic processing.

Time/Horizon Attributes

Coherence

A post-stack attribute that measures the continuity between seismic traces in a specified window along a picked horizon. It can be used to map the lateral extent of a formation. It can also be used to see faults, channels or other discontinuous features.

Although it should be used along a specified horizon, many software packages compute this attribute along arbitrary time-slices.

Dip

A post-stack attribute that computes, for each trace, the best fit plane (3D) or line (2D) between its immediate neighbor traces on a horizon and outputs the magnitude of dip (gradient) of said plane or line measured in degrees. This can be used to create a pseudo paleogeologic map on a horizon slice.

Azimuth

A post-stack attribute that computes, for each trace, the best fit plane (3D) between its immediate neighbor traces on a horizon and outputs the direction of maximum slope (dip direction) measured in degrees, clockwise from north. This is not to be confused with the geological concept of azimuth, which is equivalent to strike and is measured 90° counterclockwise from the dip direction.

Curvature

A group of post-stack attributes that are computed from the curvature of a specified horizon. These attributes include: magnitude or direction of maximum curvature, magnitude or direction of minimum curvature, magnitude of curvature along the horizon's azimuth (dip) direction, magnitude of curvature along the horizon's strike direction, magnitude of curvature of a contour line along a horizon.

Frequency Attributes

These attributes involve separating and classifying seismic events within each trace based on their frequency content. The application of these attributes is commonly called spectral decomposition. The starting point of spectral decomposition is to decompose each 1D trace from the time domain into its corresponding 2D representation in the time-frequency domain by means of any method of time-frequency decomposition such as: short-time Fourier transform, continuous wavelet transform, Wigner-Ville distribution, matching pursuit, among many others. Once each trace has been transformed into the time-frequency domain, a bandpass filter can be applied to view the amplitudes of seismic data at any frequency or range of frequencies.

Technically, each individual frequency or band of frequencies could be considered an attribute. The seismic data is usually filtered at various frequency ranges in order to show certain geological patterns that may not be obvious in the other frequency bands. There is an inverse relationship between the thickness of a rock layer and the corresponding peak frequency of its seismic reflection. That is, thinner rock layers are much more apparent at higher frequencies and thicker rock layers are much more apparent at lower frequencies. This can be used to qualitatively identify thinning or thickening of a rock unit in different directions.

Spectral decomposition has also been widely used as a direct hydrocarbon indicator.

References

- Sheriff, R. E., Geldart, L. P. (1995). Exploration Seismology (2nd ed.). Cambridge University Press. pp. 209–210. ISBN 0-521-46826-4.

- Sheriff, R. E., Geldart, L. P. (1995). Exploration Seismology (2nd ed.). Cambridge University Press. p. 200. ISBN 0-521-46826-4.

- Sheriff, R. E., Geldart, L. P. (1995). Exploration Seismology (2nd ed.). Cambridge University Press. p. 260. ISBN 0-521-46826-4.

- Sheriff, R. E., Geldart, L. P. (1995). Exploration Seismology (2nd ed.). Cambridge University Press. p. 292. ISBN 0-521-46826-4.

- Sheriff, R. E., Geldart, L. P. (1995). Exploration Seismology (2nd ed.). Cambridge University Press. p. 349. ISBN 0-521-46826-4.

- Sheriff, R. E.; Geldart, L. P. (1995). Exploration Seismology (2nd ed.). ISBN 9781139643115.

- Sheriff, R.E. (2002). Encyclopedic Dictionary of Applied Geophysics (4 ed.). Society of Exploration Geophysicists. ISBN 1-56080-118-2.

- Castagna, J.P.; Backus, M.M. (1993). Offset Dependent Reflectivity - Theory and Practice of AVO Analysis. Society of Exploration Geophysicists. ISBN 1-56080-059-3.

- Long, A. (October–November 2004). "What is Wave Equation Pre-Stack Depth Migration? An Overview" (pdf). PESA News. Retrieved 24 October 2015.

- Milkereit, B.; Eaton, D.; Salisbury, M.; Adam, E.; Bohlen, Thomas (2003). "3D Seismic Imaging for Mineral Exploration" (PDF). Commission on Controlled-Source Seismology: Deep Seismic Methods. Retrieved September 8, 2013.

- Louie, John N.; Pullammanappallil, S. K. (2011). "Advanced seismic imaging for geothermal development" (PDF). New Zealand Geothermal Workshop 2011 Proceedings. Retrieved

September 8, 2013.

- Barley, B. & Summers, T. (2007). "Multi-azimuth and wide-azimuth seismic: Shallow to deep water, exploration to production". The Leading Edge. SEG. 26 (4): 450–458. doi:10.1190/1.2723209. Retrieved September 8, 2013.

- Mike Howard (2007). "Marine seismic surveys with enhanced azimuth coverage: Lessons in survey design and acquisition" (PDF). The Leading Edge. SEG. 26 (4): 480. doi:10.1190/1.2723212. Retrieved September 8, 2013.

- Barley, B. & Summers, T. (2007). "Multi-azimuth and wide-azimuth seismic: Shallow to deep water, exploration to production". The Leading Edge. SEG. 26 (4): 456–457. doi:10.1190/1.2723209. Retrieved September 8, 2013.

- Jon Cocker (2011). "Land 3-D Seismic Survey Designed To Meet New Objectives". E & P. Hart Energy. Retrieved 12 March 2012.s

- Gausland, I. (2000). "Impact of seismic surveys on marine life" (PDF). The Leading Edge. SEG: 904. Retrieved 8 March 2012.

- McCauley, R.D.; et al. (2000). "Marine seismic surveys: A study of environmental implications" (PDF). APPEA: 692–708. Retrieved 8 March 2012.

- Beaudoin, G. (2010). "Imaging the invisible — BP's path to OBS nodes". SEG Expanded Abstracts. SEG. 29: 3734. doi:10.1190/1.3513626.

- Grusic, V., and Orlic, M., Early Observations of Rotor Clouds by Andrija Mohorovičić, Bulletin of the American Meteorlogical Society, May 2007, pp. 693-700, accessed 4 January 2010.

- Beaudoin, G.; Reasnor, M. (2010). "Atlantis time-lapse ocean bottom node survey: a project team's journey from acquisition through processing". SEG Expanded Abstracts. SEG (29): 4155.

Understanding Structure of the Earth

The structure of the Earth is spherical and has an inner core and an outer core. The crust of the Earth is the outer solid shell which is generated by igneous processes. The chapter strategically encompasses and incorporates the basic understanding of the structure of the Earth.

Structure of the Earth

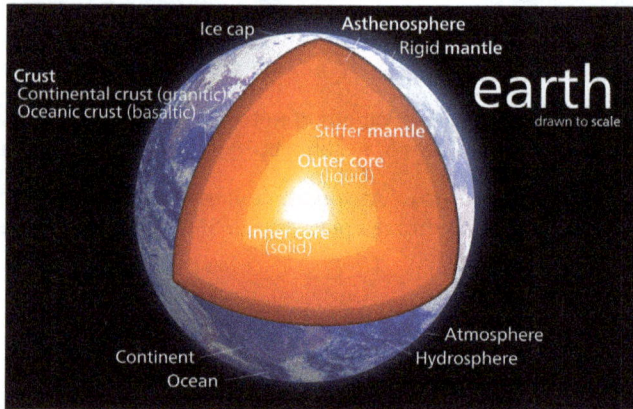

Structure of the Earth

The interior structure of the Earth is layered in spherical shells, like an onion. These layers can be defined by their chemical and their rheological properties. Earth has an outer silicate solid crust, a highly viscous mantle, a liquid outer core that is much less viscous than the mantle, and a solid inner core. Scientific understanding of the internal structure of the Earth is based on observations of topography and bathymetry, observations of rock in outcrop, samples brought to the surface from greater depths by volcanoes or volcanic activity, analysis of the seismic waves that pass through the Earth, measurements of the gravitational and magnetic fields of the Earth, and experiments with crystalline solids at pressures and temperatures characteristic of the Earth's deep interior.

Mass

The force exerted by Earth's gravity can be used to calculate its mass. Astronomers can also calculate Earth's mass by observing the motion of orbiting satellites. Earth's average density can be determined through gravitometric experiments, which have historically involved pendulums.

The mass of Earth is about 6×10^{24} kg.

Structure

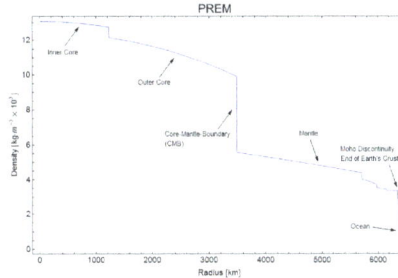

Earth's radial density distribution according to the preliminary reference earth model (PREM).

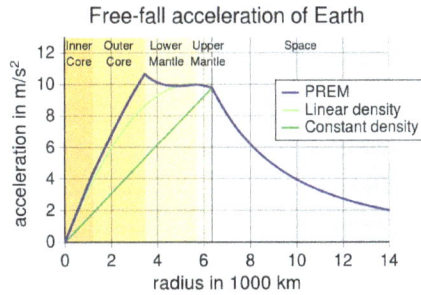

Earth's gravity according to the preliminary reference earth model (PREM). Comparison to approximations using constant and linear density for Earth's interior.

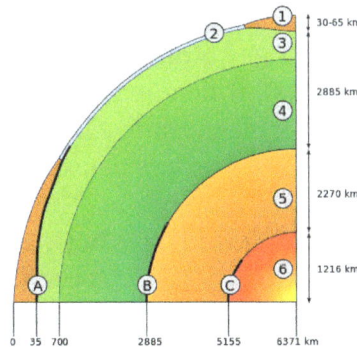

Schematic view of the interior of Earth. 1. continental crust – 2. oceanic crust – 3. upper mantle – 4. lower mantle – 5. outer core – 6. inner core – A: Mohorovičić discontinuity – B: Gutenberg Discontinuity – C: Lehmann–Bullen discontinuity.

The structure of Earth can be defined in two ways: by mechanical properties such as rheology, or chemically. Mechanically, it can be divided into lithosphere, asthenosphere, mesospheric mantle, outer core, and the inner core. The interior of Earth is divided into 5 important layers. Chemically, Earth can be divided into the crust, upper mantle, lower mantle, outer core, and inner core. The geologic component layers of Earth are at the following depths below the surface:

Depth		Layer
Kilometres	Miles	
0–60	0–37	Lithosphere (locally varies between 5 and 200 km)
0–35	0–22	... Crust (locally varies between 5 and 70 km)
35–60	22–37	... Uppermost part of mantle
35–2,890	22–1,790	Mantle
210-270	130-168	... Upper mesosphere (upper mantle)
660–2,890	410–1,790	... Lower mesosphere (lower mantle)
2,890–5,150	1,790–3,160	Outer core
5,150–6,360	3,160–3,954	Inner core

The layering of Earth has been inferred indirectly using the time of travel of refracted and reflected seismic waves created by earthquakes. The core does not allow shear waves to pass through it, while the speed of travel (seismic velocity) is different in other layers. The changes in seismic velocity between different layers causes refraction owing to Snell's law, like light bending as it passes through a prism. Likewise, reflections are caused by a large increase in seismic velocity and are similar to light reflecting from a mirror.

Crust

The crust ranges from 5–70 km (~3–44 miles) in depth and is the outermost layer. The thin parts are the oceanic crust, which underlie the ocean basins (5–10 km) and are composed of dense (mafic) iron magnesium silicate igneous rocks, like basalt. The thicker crust is continental crust, which is less dense and composed of (felsic) sodium potassium aluminium silicate rocks, like granite. The rocks of the crust fall into two major categories – sial and sima (Suess,1831–1914). It is estimated that sima starts about 11 km below the Conrad discontinuity (a second order discontinuity). The uppermost mantle together with the crust constitutes the lithosphere. The crust-mantle boundary occurs as two physically different events. First, there is a discontinuity in the seismic velocity, which is most commonly known as the Mohorovičić discontinuity or Moho. The cause of the Moho is thought to be a change in rock composition from rocks containing plagioclase feldspar (above) to rocks that contain no feldspars (below). Second, in oceanic crust, there is a chemical discontinuity between ultramafic cumulates and tectonized harzburgites, which has been observed from deep parts of the oceanic crust that have been obducted onto the continental crust and preserved as ophiolite sequences.

Many rocks now making up Earth's crust formed less than 100 million (1×10^8) years ago; however, the oldest known mineral grains are 4.4 billion (4.4×10^9) years old, indicating that Earth has had a solid crust for at least that long.

Mantle

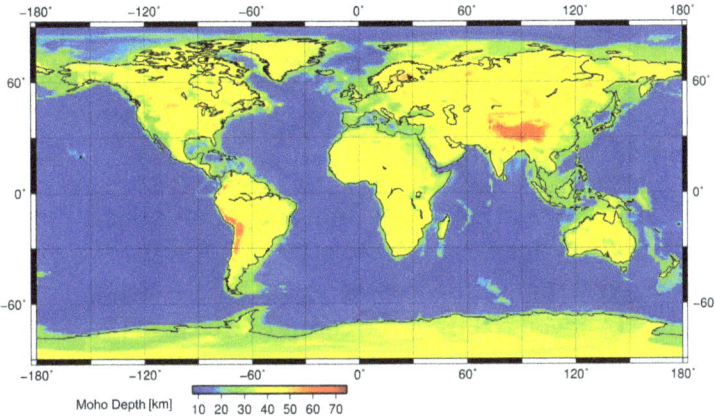

World map showing the position of the Moho.

Earth's mantle extends to a depth of 2,890 km, making it the thickest layer of Earth. The mantle is divided into upper and lower mantle. The upper and lower mantle are separated by the transition zone. The lowest part of the mantle next to the core-mantle boundary is known as the D″ (D prime prime) layer. The pressure at the bottom of the mantle is ~140 GPa (1.4 Matm). The mantle is composed of silicate rocks that are rich in iron and magnesium relative to the overlying crust. Although solid, the high temperatures within the mantle cause the silicate material to be sufficiently ductile that it can flow on very long timescales. Convection of the mantle is expressed at the surface through the motions of tectonic plates. As there is intense and increasing pressure as one travels deeper into the mantle, the lower part of the mantle flows less easily than does the upper mantle (chemical changes within the mantle may also be important). The viscosity of the mantle ranges between 10^{21} and 10^{24} Pa·s, depending on depth. In comparison, the viscosity of water is approximately 10^{-3} Pa·s and that of pitch is 10^7 Pa·s. The source of heat that drives plate tectonics is the primordial heat left over from the planet's formation as well as the radioactive decay of uranium, thorium, and potassium in Earth's crust and mantle.

Core

The average density of Earth is 5,515 kg/m³. Because the average density of surface material is only around 3,000 kg/m³, we must conclude that denser materials exist within Earth's core. Seismic measurements show that the core is divided into two parts, a "solid" inner core with a radius of ~1,220 km and a liquid outer core extending beyond it to a radius of ~3,400 km. The densities are between 9,900 and 12,200 kg/m³ in the outer core and 12,600–13,000 kg/m³ in the inner core.

The inner core was discovered in 1936 by Inge Lehmann and is generally believed to be composed primarily of iron and some nickel. It is not necessarily a solid, but, because it is able to deflect seismic waves, it must behave as a solid in some fashion. Experimental evidence has at times been critical of crystal models of the core. Other experimental studies show a discrepancy under high pressure: diamond anvil (static) studies at core pressures yield melting temperatures that are approximately 2000 K below those from shock laser (dynamic) studies. The laser studies create plasma, and the results are suggestive that constraining inner core conditions will depend on whether the inner core is a solid or is a plasma with the density of a solid. This is an area of active research.

In early stages of Earth's formation about four and a half billion (4.5×10^9) years ago, melting would have caused denser substances to sink toward the center in a process called planetary differentiation, while less-dense materials would have migrated to the crust. The core is thus believed to largely be composed of iron (80%), along with nickel and one or more light elements, whereas other dense elements, such as lead and uranium, either are too rare to be significant or tend to bind to lighter elements and thus remain in the crust. Some have argued that the inner core may be in the form of a single iron crystal.

Under laboratory conditions a sample of iron–nickel alloy was subjected to the corelike pressures by gripping it in a vise between 2 diamond tips (diamond anvil cell), and then heating to approximately 4000 K. The sample was observed with x-rays, and strongly supported the theory that Earth's inner core was made of giant crystals running north to south.

The liquid outer core surrounds the inner core and is believed to be composed of iron mixed with nickel and trace amounts of lighter elements.

Recent speculation suggests that the innermost part of the core is enriched in gold, platinum and other siderophile elements.

The matter that comprises Earth is connected in fundamental ways to matter of certain chondrite meteorites, and to matter of outer portion of the Sun. There is good reason to believe that Earth is, in the main, like a chondrite meteorite. Beginning as early as 1940, scientists, including Francis Birch, built geophysics upon the premise that Earth is like ordinary chondrites, the most common type of meteorite observed impacting Earth, while totally ignoring another, albeit less abundant type, called enstatite chondrites. The principal difference between the two meteorite types is that enstatite chondrites formed under circumstances of extremely limited available oxygen, leading to certain normally oxyphile elements existing either partially or wholly in the alloy portion that corresponds to the core of Earth.

Dynamo theory suggests that convection in the outer core, combined with the Coriolis effect, gives rise to Earth's magnetic field. The solid inner core is too hot to hold a permanent magnetic field but probably acts to stabilize the magnetic field generated by the liquid outer core. The average magnetic field strength in Earth's outer core is estimated to be 25 Gauss (2.5 mT), 50 times stronger than the magnetic field at the surface.

Recent evidence has suggested that the inner core of Earth may rotate slightly faster than the rest of the planet; however, more recent studies in 2011 found this hypothesis to be inconclusive. Options remain for the core which may be oscillatory in nature or a chaotic system. In August 2005 a team of geophysicists announced in the journal *Science* that, according to their estimates, Earth's inner core rotates approximately 0.3 to 0.5 degrees per year faster relative to the rotation of the surface.

The current scientific explanation for Earth's temperature gradient is a combination of heat left over from the planet's initial formation, decay of radioactive elements, and freezing of the inner core.

Historical Development of Alternative Conceptions

Edmond Halley's hypothesis.

In 1692 Edmond Halley (in a paper printed in *Philosophical Transactions of Royal Society of London*) put forth the idea of Earth consisting of a hollow shell about 500 miles thick, with two inner concentric shells around an innermost core, corresponding to the diameters of the planets Venus, Mars, and Mercury respectively. Halley's construct was a method of accounting for the (flawed) values of the relative density of Earth and the Moon that had been given by Sir Isaac Newton, in *Principia* (1687). "Sir Isaac Newton has demonstrated the Moon to be more solid than our Earth, as 9 to 5," Halley remarked; "why may we not then suppose four ninths of our globe to be cavity?"

Crust (Geology)

In geology, the crust is the outermost solid shell of a rocky planet or natural satellite, which is chemically distinct from the underlying mantle. The crusts of Earth, the Moon, Mercury, Venus, Mars, Io, and other planetary bodies have been generated largely by igneous processes, and these crusts are richer in incompatible elements than their respective mantles.

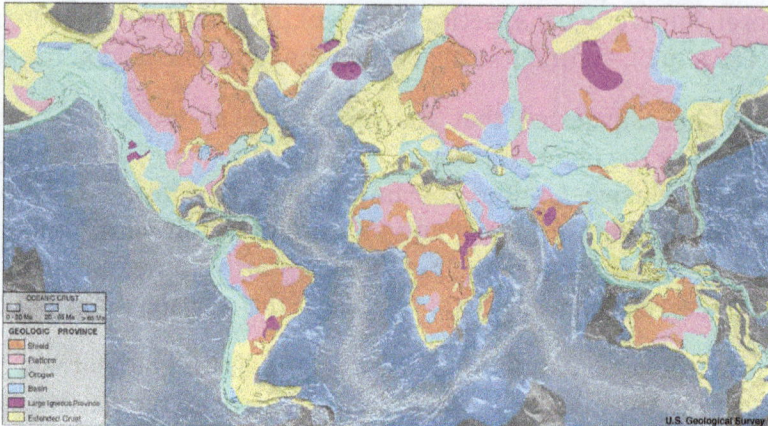

Geologic provinces of the World (USGS)

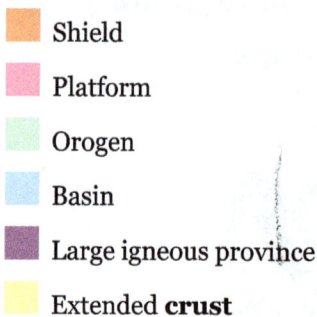

Shield

Platform

Orogen

Basin

Large igneous province

Extended **crust**

Oceanic crust:

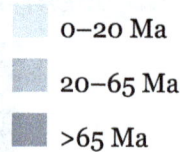

0–20 Ma

20–65 Ma

>65 Ma

Earth's Crust

The crust of the Earth is composed of a great variety of igneous, metamorphic, and sedimentary rocks. The crust is underlain by the mantle. The upper part of the mantle is composed mostly of peridotite, a rock denser than rocks common in the overlying crust. The boundary between the crust and mantle is conventionally placed at the Mohorovičić discontinuity, a boundary defined by a contrast in seismic velocity. The crust occupies less than 1% of Earth's volume.

The oceanic crust of the sheet is different from its continental crust.

- The oceanic crust is 5 km (3 mi) to 10 km (6 mi) thick and is composed primarily of basalt, diabase, and gabbro.

- The continental crust is typically from 30 km (20 mi) to 50 km (30 mi) thick and is mostly composed of slightly less dense rocks than those of the oceanic crust. Some of these less dense rocks, such as granite, are common in the continental crust but rare to absent in the oceanic crust.

Both the continental and oceanic crust "float" on the mantle. Because the continental crust is thicker, it extends both to greater elevations and greater depth than the oceanic crust. The slightly lower density of felsic continental rock compared to basaltic oceanic rock contributes to the higher relative elevation of the top of the continental crust. As the top of the continental crust reaches elevations higher than that of the oceanic, water runs off the continents and collects above the oceanic crust. Because of the change in velocity of seismic waves it is believed that beneath continents at a certain depth continental crust (sial) becomes close in its physical properties to oceanic crust (sima), and the transition zone is referred to as the Conrad discontinuity.

The temperature of the crust increases with depth, reaching values typically in the range from about 200 °C (392 °F) to 400 °C (752 °F) at the boundary with the underlying mantle. The crust and underlying relatively rigid uppermost mantle make up the lithosphere. Because of convection in the underlying plastic (although non-molten) upper mantle and asthenosphere, the lithosphere is broken into tectonic plates that move. The temperature increases by as much as 30 °C (about 50 °F) for every kilometer locally in the upper part of the crust, but the geothermal gradient is smaller in deeper crust.

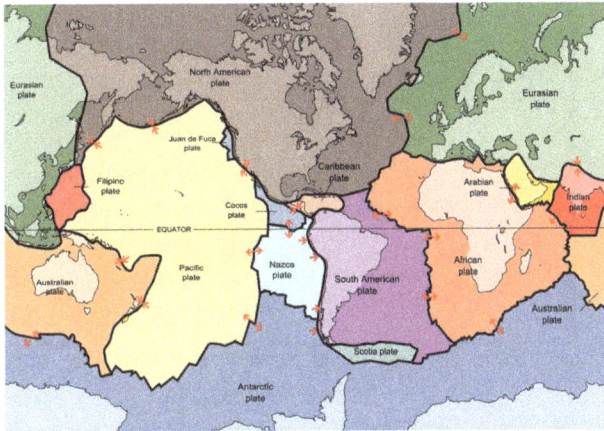

Plates in the crust of Earth

Partly by analogy to what is known about the Moon, Earth is considered to have differentiated from an aggregate of planetesimals into its core, mantle and crust within about 100 million years of the formation of the planet, 4.6 billion years ago. The primordial crust was very thin and was probably recycled by much more vigorous plate tectonics and destroyed by significant asteroid impacts, which were much more common in the early stages of the solar system.

Earth has probably always had some form of basaltic crust, but the age of the oldest oceanic crust today is only about 200 million years. In contrast, the bulk of the continental crust is much older. The oldest continental crustal rocks on Earth have ages in the range from about 3.7 to 4.28 billion years and have been found in the Narryer Gneiss Terrane in Western Australia, in the Acasta Gneiss in the Northwest Territories on the Canadian Shield, and on other cratonic regions such as those on the Fennoscandian Shield. Some zircon with age as great as 4.3 billion years has been found in the Narryer Gneiss Terrane.

The average age of the current Earth's continental crust has been estimated to be about 2.0 billion years. Most crustal rocks formed before 2.5 billion years ago are located in cratons. Such old continental crust and the underlying mantle asthenosphere are less dense than elsewhere in Earth and so are not readily destroyed by subduction. Formation of new continental crust is linked to periods of intense orogeny; these periods coincide with the formation of the supercontinents such as Rodinia, Pangaea and Gondwana. The crust forms in part by aggregation of island arcs including granite and metamorphic fold belts, and it is preserved in part by depletion of the underlying mantle to form buoyant lithospheric mantle.

Composition

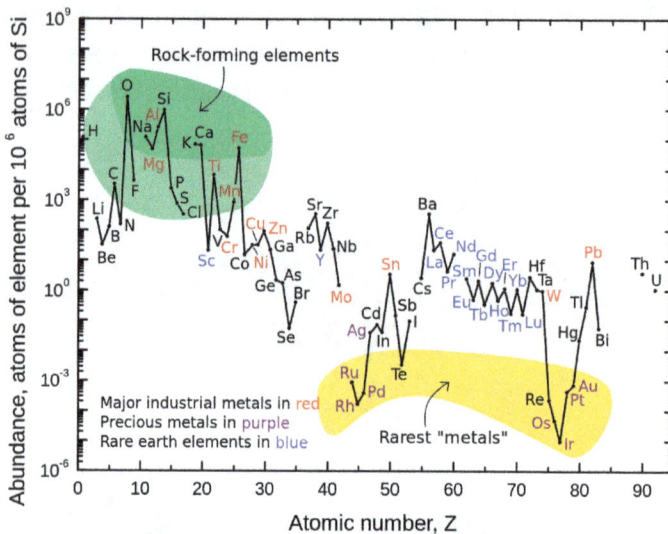

Abundance (atom fraction) of the chemical elements in Earth's upper continental crust as a function of atomic number. The rarest elements in the crust (shown in yellow) are not the heaviest, but are rather the siderophile (iron-loving) elements in the Goldschmidt classification of elements. These have been depleted by being relocated deeper into Earth's core. Their abundance in meteoroid materials is higher. Additionally, tellurium and selenium have been depleted from the crust due to formation of volatile hydrides.

The continental crust has an average composition similar to that of andesite. Continental crust is enriched in incompatible elements compared to the basaltic ocean crust and much enriched compared to the underlying mantle. Although the continental crust comprises only about 0.6 weight percent of the silicate on Earth, it contains 20% to 70% of the incompatible elements.

Most Abundant Elements of Earth's Crust	Approximate % by weight
O	46.6
Si	27.7
Al	8.1
Fe	5.0

Ca	3.6
Na	2.8
K	2.6
Mg	1.5

Oxide	Percent
SiO_2	60.6
Al_2O_3	15.9
CaO	6.4
MgO	4.7
Na_2O	3.1
Fe as FeO	6.7
K_2O	1.8
TiO_2	0.7
P_2O_5	0.1

All the other constituents except water occur only in very small quantities and total less than 1%. Estimates of average density for the upper crust range between 2.69 and 2.74 g/cm^3 and for lower crust between 3.0 and 3.25 g/cm^3.

Moon's Crust

A theoretical protoplanet named "Theia" is thought to have collided with the forming Earth, and part of the material ejected into space by the collision accreted to form the Moon. As the Moon formed, the outer part of it is thought to have been molten, a "lunar magma ocean." Plagioclase feldspar crystallized in large amounts from this magma ocean and floated toward the surface. The cumulate rocks form much of the crust. The upper part of the crust probably averages about 88% plagioclase (near the lower limit of 90% defined for anorthosite): the lower part of the crust may contain a higher percentage of ferromagnesian minerals such as the pyroxenes and olivine, but even that lower part probably averages about 78% plagioclase. The underlying mantle is denser and olivine-rich.

The thickness of the crust ranges between about 20 and 120 km. Crust on the far side of the Moon averages about 12 km thicker than that on the near side. Estimates of average thickness fall in the range from about 50 to 60 km. Most of this plagioclase-rich crust formed shortly after formation of the moon, between about 4.5 and 4.3 billion years ago. Perhaps 10% or less of the crust consists of igneous rock added after the formation of the initial plagioclase-rich material. The best-characterized and most voluminous of these later additions are the mare basalts formed between about 3.9 and 3.2 billion years ago. Minor volcanism continued after 3.2 billion years, perhaps as recently as 1 billion years ago. There is no evidence of plate tectonics.

Study of the Moon has established that a crust can form on a rocky planetary body significantly smaller than Earth. Although the radius of the Moon is only about a quarter that of Earth, the lunar crust has a significantly greater average thickness. This thick crust formed almost immediately after formation of the Moon. Magmatism continued after the period of intense meteorite impacts ended about 3.9 billion years ago, but igneous rocks younger than 3.9 billion years make up only a minor part of the crust.

Types of Crust

Continental Crust

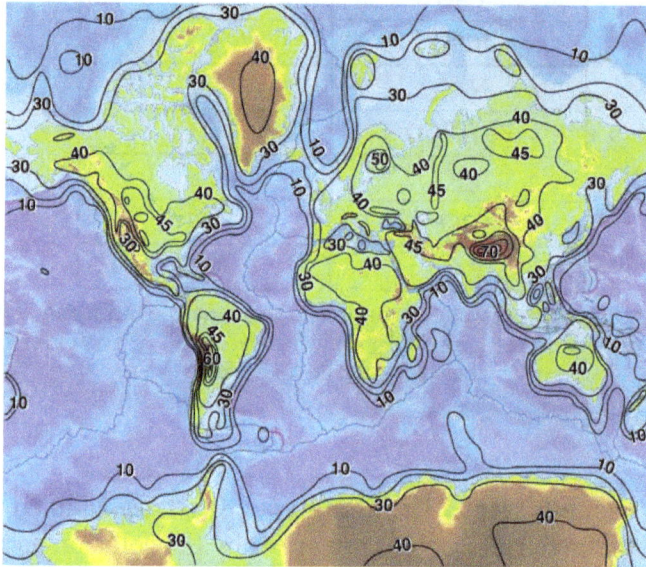

The thickness of Earth's crust (km)

The continental crust is the layer of igneous, sedimentary, and metamorphic rocks that forms the continents and the areas of shallow seabed close to their shores, known as continental shelves. This layer is sometimes called *sial* because its bulk composition is more felsic compared to the oceanic crust, called *sima* which has a more mafic bulk composition. Changes in seismic wave velocities have shown that at a certain depth (the Conrad discontinuity), there is a reasonably sharp contrast between the more felsic upper continental crust and the lower continental crust, which is more mafic in character.

The continental crust consists of various layers, with a bulk composition that is intermediate to felsic. The average density of continental crust is about 2.7 g/cm³, less dense than the ultramafic material that makes up the mantle, which has a density of around 3.3 g/cm³. Continental crust is also less dense than oceanic crust, whose density is about 2.9 g/cm³. At 25 to 70 km, continental crust is considerably thicker than oceanic crust, which has an average thickness of around 7–10 km. About 40% of Earth's surface is currently occupied by continental crust. It makes up about 70% of the volume of Earth's crust.

Importance

Because the surface of continental crust mainly lies above sea level, its existence allowed land life to evolve from marine life. Its existence also provides broad expanses of shallow water known as epeiric seas and continental shelves where complex metazoan life could become established during early Paleozoic time, in what is now called the Cambrian explosion.

Origin

There is little evidence of continental crust prior to 3.5 Ga, and there was relatively rapid development on shield areas consisting of continental crust between 3.0 and 2.5 Ga. All continental crust ultimately derives from the fractional differentiation of oceanic crust over many eons. This process has been and continues today primarily as a result of the volcanism associated with subduction.

Forces at Work

In contrast to the persistence of continental crust, the size, shape, and number of continents are constantly changing through geologic time. Different tracts rift apart, collide and recoalesce as part of a grand supercontinent cycle. There are currently about 7 billion cubic kilometers of continental crust, but this quantity varies because of the nature of the forces involved. The relative permanence of continental crust contrasts with the short life of oceanic crust. Because continental crust is less dense than oceanic crust, when active margins of the two meet in subduction zones, the oceanic crust is typically subducted back into the mantle. Continental crust is rarely subducted (this may occur where continental crustal blocks collide and over-thicken, causing deep melting under mountain belts such as the Himalayas or the Alps). For this reason the oldest rocks on Earth are within the cratons or cores of the continents, rather than in repeatedly recycled oceanic crust; the oldest intact crustal fragment is the Acasta Gneiss at 4.01 Ga, whereas the oldest oceanic crust (located on the Pacific Plate offshore of Kamchatka) is from the Jurassic (~180 Ma). Continental crust and the rock layers that lie on and within it are thus the best archive of Earth's history.

The height of mountain ranges is usually related to the thickness of crust. This results from the isostasy associated with orogeny (mountain formation). The crust is thickened by the compressive forces related to subduction or continental collision. The buoyancy of the crust forces it upwards, the forces of the collisional stress balanced by gravity and erosion. This forms a keel or mountain root beneath the mountain range, which is where the thickest crust is found. The thinnest continental crust is found in rift zones, where the crust is thinned by detachment faulting and eventually severed, replaced by oceanic crust. The edges of continental fragments formed this way (both sides of the Atlantic Ocean, for example) are termed passive margins.

The high temperatures and pressures at depth, often combined with a long history of complex distortion, cause much of the lower continental crust to be metamorphic - the main exception to this being recent igneous intrusions. Igneous rock may also be "underplated" to the underside of the crust, i.e. adding to the crust by forming a layer immediately beneath it.

Continental crust is produced and (far less often) destroyed mostly by plate tectonic processes, especially at convergent plate boundaries. Additionally, continental crustal material is transferred to oceanic crust by sedimentation. New material can be added to the continents by the partial melting of oceanic crust at subduction zones, causing the lighter material to rise as magma, forming volcanoes. Also, material can be accreted horizontally when volcanic island arcs, seamounts or similar structures collide with the side of the continent as a result of plate tectonic movements. Continental crust is also lost through erosion and sediment subduction, tectonic erosion of forearcs, delamination, and deep subduction of continental crust in collision zones. Many theories of crustal growth are controversial, including rates of crustal growth and recycling, whether the lower crust is recycled differently from the upper crust, and over how much of Earth history plate tectonics has operated and so could be the dominant mode of continental crust formation and destruction.

It is a matter of debate whether the amount of continental crust has been increasing, decreasing, or remaining constant over geological time. One model indicates that at prior to 3.7 Ga ago continental crust constituted less than 10% of the present amount. By 3.0 Ga ago the amount was about 25%, and following a period of rapid crustal evolution it was about 60% of the current amount by 2.6 Ga ago. The growth of continental crust appears to have occurred in *spurts* of increased activity corresponding to five episodes of increased production through geologic time.

Oceanic Crust

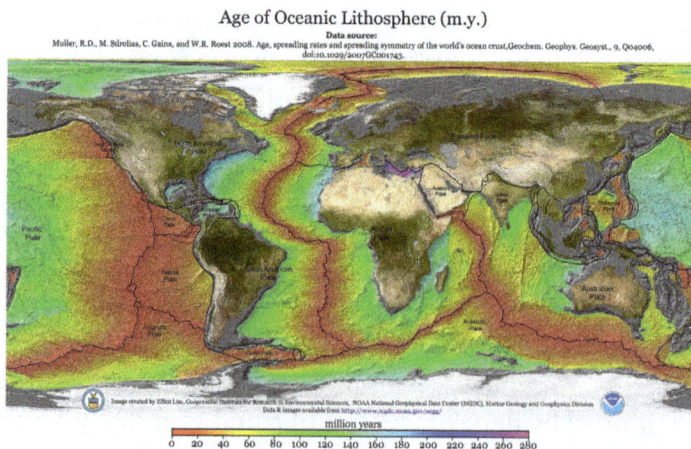

Age of Oceanic Lithosphere (m.y.)
Data source:
Muller, R.D., M. Sdrolias, C. Gaina, and W.R. Roest 2008. Age, spreading rates and spreading symmetry of the world's ocean crust,Geochem. Geophys. Geosyst., 9, Q04006, doi:10.1029/2007GC001743.

million years
0   20   40   60   80   100   120   140   160   180   200   220   240   260   280

Colors indicate the age of oceanic lithosphere, wherein red indicates the youngest age, and blue indicates the oldest age. The lines represent tectonic plates.

Oceanic crust is the uppermost layer of the oceanic portion of a tectonic plate. The crust overlies the solidified and uppermost layer of the mantle. The crust and the solid mantle layer together constitute oceanic lithosphere.

Oceanic crust is the result of erupted mantle material originating from below the plate, cooled and in most instances, modified chemically by seawater. This occurs mostly at mid-ocean ridges, but also at scattered hotspots, and also in rare but powerful occurrences known as flood basalt eruptions. It is primarily composed of mafic rocks, or sima, which is rich in iron and magnesium. It is thinner than continental crust, or sial, generally less than 10 kilometers thick; however it is denser, having a mean density of about 2.9 grams per cubic centimeter as opposed to continental crust which has a density of about 2.7 grams per cubic centimeter.

Composition

Although a complete section of oceanic crust has not yet been drilled, geologists have several pieces of evidence that help them understand the ocean floor. Estimations of composition are based on analyses of ophiolites (sections of oceanic crust that are preserved on the continents), comparisons of the seismic structure of the oceanic crust with laboratory determinations of seismic velocities in known rock types, and samples recovered from the ocean floor by submersibles, dredging (especially from ridge crests and fracture zones) and drilling. Oceanic crust is significantly simpler than continental crust and generally can be divided in three layers.

- Layer 1 is on an average 0.4 km thick. It consists of unconsolidated or semiconsolidated sediments, usually thin or even not present near the mid-ocean ridges but thickens farther away from the ridge. Near the continental margins sediment is terrigenous, meaning derived from the land, unlike deep sea sediments which are made of tiny shells of marine organisms, usually calcareous and siliceous, or it can be made of volcanic ash and terrigenous sediments transported by turbidity currents.

- Layer 2 could be divided into two parts: layer 2A – 0.5 km thick uppermost volcanic layer of glassy to finely crystalline basalt usually in the form of pillow basalt, and layer 2B – 1.5 km thick layer composed of diabase dikes.

- Layer 3 is formed by slow cooling of magma beneath the surface and consists of coarse grained gabbros and cumulate ultramafic rocks. It constitutes over two-thirds of oceanic crust volume with almost 5 km thickness.

Geochemistry

The most voluminous volcanic rocks of the ocean floor are the mid-oceanic ridge basalts, which are derived from low-potassium tholeiitic magmas. These rocks have low concentrations of large ion lithophile elements (LILE), light rare earth elements (LREE), volatile elements and other highly incompatible elements. There can be found basalts enriched with incompatible elements, but they are rare and associated with mid-ocean ridge hot spots such as surroundings of Galapagos Islands, the Azores and Iceland.

Life Cycle

Oceanic crust is continuously being created at mid-ocean ridges. As plates diverge at these ridges, magma rises into the upper mantle and crust. As it moves away from the ridge, the lithosphere becomes cooler and denser, and sediment gradually builds on top of it. The youngest oceanic lithosphere is at the oceanic ridges, and it gets progressively older away from the ridges.

As the mantle rises it cools and melts, as the pressure decreases and it crosses the solidus. The amount of melt produced depends only on the temperature of the mantle as it rises. Hence most oceanic crust is the same thickness (7 ± 1 km). Very slow spreading ridges (<1 cm·yr^{-1} half-rate) produce thinner crust (4–5 km thick) as the mantle has a chance to cool on upwelling and so it crosses the solidus and melts at lesser depth, thereby producing less melt and thinner crust. An example of this is the Gakkel Ridge under the Arctic Ocean. Thicker than average crust is found above plumes as the mantle is hotter and hence it crosses the solidus and melts at a greater depth, creating more melt and a thicker crust. An example of this is Iceland which has crust of thickness ~20 km.

The oceanic lithosphere subducts at what are known as convergent boundaries. These boundaries can exist between oceanic lithosphere on one plate and oceanic lithosphere on another, or between oceanic lithosphere on one plate and continental lithosphere on another. In the second situation, the oceanic lithosphere always subducts because the continental lithosphere is less dense. The subduction process consumes older oceanic lithosphere, so oceanic crust is seldom more than 200 million years old. The process of super-continent formation and destruction via repeated cycles of creation and destruction of oceanic crust is known as the Wilson cycle.

The oldest large scale oceanic crust is in the west Pacific and north-west Atlantic - both are about up to 180-200 million years old. However, parts of the eastern Mediterranean Sea are remnants of the much older Tethys ocean, at about 270 and up to 340 million years old.

Magnetic Anomalies

The *oceanic crust* displays an interesting pattern of parallel magnetic lines, parallel to the ocean ridges, frozen in the basalt. In the 1950s, scientists mapped the magnetic field generated by rocks on the ocean floor. They noticed a symmetrical pattern of positive and negative magnetic lines as they moved along the ocean floor, and the line of symmetry was at the mid ocean ridge. That the anomalies were symmetrical at the mid-ocean ridge was explained by the hypothesis that new rock was being formed by magma at the mid-ocean ridges, and the ocean floor was spreading out from this point. When the magma cooled to form rock, it aligned itself with the current position of the north magnetic pole of the Earth (which has reversed many times in its past) at the time of its cooling. New magma forced the older cooled magma away from the ridge. Approximately half of the new rock was formed on one side of the ridge and half on the other.

Mantle (Geology)

The mantle is a layer inside a terrestrial planet and some other rocky planetary bodies. For a mantle to form, the planetary body must be large enough to have undergone the process of planetary differentiation by density. The mantle lies between the core below and the crust above. The terrestrial planets (Earth, Venus, Mars and Mercury), the Moon, two of Jupiter's moons (Io and Europa) and the asteroid Vesta each have a mantle made of silicate rock. Interpretation of spacecraft data suggests that at least two other moons of Jupiter (Ganymede and Callisto), as well as Titan and Triton each have a mantle made of ice or other solid volatile substances.

Earth's Mantle

The interior of Earth, similar to the other terrestrial planets, is chemically divided into layers. The mantle is a layer between the crust and the outer core. Earth's mantle is a silicate rocky shell with an average thickness of 2,886 kilometres (1,793 mi). The mantle makes up about 84% of Earth's volume. It is predominantly solid but in geological time it behaves as a very viscous fluid. The mantle encloses the hot core rich in iron and nickel, which makes up about 15% of Earth's volume. Past episodes of melting and volcanism at the shallower levels of the mantle have produced a thin crust of crystallized melt products near the surface. Information about the structure and composition of the mantle has been obtained from geophysical investigation and from direct geoscientific analyses of Earth mantle-derived xenoliths and mantle that has been exposed by mid-oceanic ridge spreading.

Two main zones are distinguished in the upper mantle: the inner asthenosphere composed of plastic flowing rock of varying thickness, on average about 200 km (120 mi) thick, and the lowermost part of the lithosphere composed of rigid rock about 50 to 120 km (31 to 75 mi) thick. A thin crust, the upper part of the lithosphere, surrounds the mantle and is about 5 to 75 km (3.1 to 46.6 mi) thick. Recent analysis of hydrous ringwoodite from the mantle suggests that there is between one and three times as much water in the transition zone between the lower and upper mantle than in all the world's oceans combined.

In some places under the ocean the mantle is actually exposed on the surface of Earth. There are also a few places on land where mantle rock has been pushed to the surface by tectonic activity, most notably the Tablelands region of Gros Morne National Park in the Canadian province of Newfoundland and Labrador and St. John's Island, Egypt or Zabargad in the Red Sea. (Also Troodos Ophiolite, Lizard complex, Semail Ophiolite, and other Ophiolites)

Structure

The mantle is divided into sections which are based upon results from seismology. These layers (and their thicknesses/depths) are the following: the upper mantle (starting at the Moho, or base of the crust around 7 to 35 km (4.3 to 21.7 mi) downward to 410 km (250 mi)), the transition zone (410–660 km or 250–410 mi), the lower mantle (660–2,891 km or 410–1,796 mi), and anomalous core–mantle boundary with a variable thickness (on average ~200 km (120 mi) thick).

The top of the mantle is defined by a sudden increase in seismic velocity, which was first noted by Andrija Mohorovičić in 1909; this boundary is now referred to as the Mohorovičić discontinuity or "Moho". The uppermost mantle plus overlying crust are relatively rigid and form the lithosphere, an irregular layer with a maximum thickness of perhaps 200 km (120 mi). Below the lithosphere the upper mantle becomes notably more plastic. In some regions below the lithosphere, the seismic shear velocity is reduced; this so-called low-velocity zone (LVZ) extends down to a depth of several hundred km. Inge Lehmann discovered a seismic discontinuity at about 220 km (140 mi) depth; although this discontinuity has been found in other studies, it is not known whether the discontinuity is ubiquitous. The transition zone is an area of great complexity; it physically separates the upper and lower mantle. Very little is known about the lower mantle apart from that it appears to be relatively seismically homogeneous. The D" layer at the core–mantle boundary separates the mantle from the core. In 2015, research using gravitational data from GRACE satellites and the long wavelength nonhydrostatic geoid indicated viscosity increases by a factor of ten to 150 about 1,000 kilometres (620 mi) below earth's surface; separate research also indicates sinking tectonic plates stall at this depth, leading Robert van der Hilst to speculate "In term's of structure and dynamics, 1,000 kilometers could be more important" (than the currently accepted 660 km depth upper–lower division). The lower mantle also contains some discontinueous zones, called "thermochemical piles" which have been interpreted as either thermally differentiated, upwellings bringing warmer material towards the surface, or as chemically differentiated material. A principal source of the heat that drives plate tectonics is the radioactive decay of uranium, thorium, and potassium in Earth's crust and mantle.

Characteristics

The mantle differs substantially from the crust in its mechanical properties which is the direct consequence of chemical composition change (expressed as different mineralogy). The distinction between crust and mantle is based on chemistry, rock types, rheology and seismic characteristics. The crust is a solidification product of mantle derived melts, expressed as various degrees of partial melting products during geologic time. Partial melting of mantle material is believed to cause incompatible elements to separate from the mantle, with less dense material floating upward through pore spaces, cracks, or fissures, that would subsequently cool and solidify at the surface. Typical mantle rocks have a higher magnesium to iron ratio and a smaller proportion of silicon and aluminium than the crust. This behavior is also predicted by experiments that partly melt rocks thought to be representative of Earth's mantle.

Mantle rocks shallower than about 410 km (250 mi) depth consist mostly of olivine, pyroxenes, spinel-structure minerals, and garnet; typical rock types are thought to be peridotite, dunite (olivine-rich peridotite), and eclogite. Between about 400 km (250 mi) and 650 km (400 mi) depth, olivine is not stable and is replaced by high pressure polymorphs with approximately the same composition: one polymorph is wadsleyite (also called *beta-spinel* type), and the other is ringwoodite (a mineral with the *gamma-spinel* structure). Below about 650 km (400 mi), all of the minerals of the upper mantle begin to become unstable. The most abundant minerals present, the silicate perovskites, have structures (but not compositions) like that of the mineral perovskite followed by the magnesium/iron oxide ferropericlase. The changes in mineralogy at about 400 and 650 km (250 and 400 mi) yield distinctive signatures in seismic records of the Earth's interior, and like the moho, are readily detected using seismic waves. These changes in mineralogy may influence mantle convection, as they result in density changes and they may absorb or release latent heat as well as depress or elevate the depth of the polymorphic phase transitions for regions of different temperatures. The changes in mineralogy with depth have been investigated by laboratory experiments that duplicate high mantle pressures, such as those using the diamond anvil.

Composition of Earth's mantle in weight percent			
Element	Amount	Compound	Amount
O	44.8		
Mg	22.8	SiO_2	46
Si	21.5	MgO	37.8
Fe	5.8	FeO	7.5
Ca	2.3	Al_2O_3	4.2
Al	2.2	CaO	3.2
Na	0.3	Na_2O	0.4
K	0.03	K_2O	0.04
Sum	99.7	Sum	99.1

The inner core is solid, the outer core is liquid, and the mantle solid/plastic. This is because of the relative melting points of the different layers (nickel–iron core, silicate crust and mantle) and the increase in temperature and pressure as depth increases. At the surface both nickel–iron alloys and silicates are sufficiently cool to be solid. In the upper mantle, the silicates are generally solid (localised regions with small amounts of melt exist); however, as the upper mantle is both hot and under relatively little pressure, the rock in the upper mantle has a relatively low viscosity. In contrast, the lower mantle is under tremendous pressure and therefore has a higher viscosity than the upper mantle. The metallic nickel–iron outer core is liquid because of the high temperature, despite the high pressure. As the pressure increases, the nickel–iron inner core becomes solid because the melting point of iron increases dramatically at these Temperature

In the mantle, temperatures range between 500 to 900 °C (932 to 1,652 °F) at the upper boundary high pressures.

with the crust; to over 4,000 °C (7,230 °F) at the boundary with the core. Although the higher temperatures far exceed the melting points of the mantle rocks at the surface (about 1200 °C for representative peridotite), the mantle is almost exclusively solid. The enormous lithostatic pressure exerted on the mantle prevents melting, because the temperature at which melting begins (the solidus) increases with pressure.

Movement

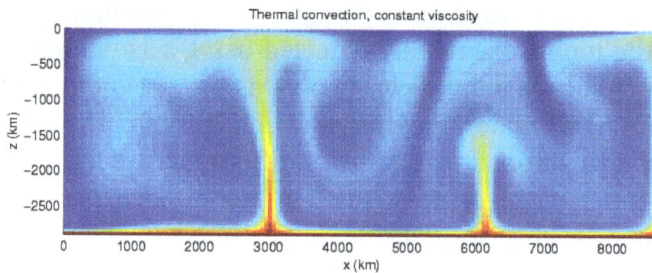

This figure is a snapshot of one time-step in a model of mantle convection. Colors closer to red are hot areas and colors closer to blue are cold areas. In this figure, heat received at the core–mantle boundary results in thermal expansion of the material at the bottom of the model, reducing its density and causing it to send plumes of hot material upwards. Likewise, cooling of material at the surface results in its sinking.

Because of the temperature difference between the Earth's surface and outer core and the ability of the crystalline rocks at high pressure and temperature to undergo slow, creeping, viscous-like deformation over millions of years, there is a convective material circulation in the mantle. Hot material upwells, while cooler (and heavier) material sinks downward. Downward motion of material occurs at convergent plate boundaries called subduction zones. Locations on the surface that lie over plumes are predicted to have high elevation (because of the buoyancy of the hotter, less-dense plume beneath) and to exhibit hot spot volcanism. The volcanism often attributed to deep mantle plumes is alternatively explained by passive extension of the crust, permitting magma to leak to the surface (the "Plate" hypothesis).

The convection of the Earth's mantle is a chaotic process (in the sense of fluid dynamics), which is thought to be an integral part of the motion of plates. Plate motion should not be confused with continental drift which applies purely to the movement of the crustal components of the continents. The movements of the lithosphere and the underlying mantle are coupled since descending lithosphere is an essential component of convection in the mantle. The observed continental drift is a complicated relationship between the forces causing oceanic lithosphere to sink and the movements within Earth's mantle.

Although there is a tendency to larger viscosity at greater depth, this relation is far from linear and shows layers with dramatically decreased viscosity, in particular in the upper mantle and at the boundary with the core. The mantle within about 200 km (120 mi) above the core–mantle boundary appears to have distinctly different seismic properties than the mantle at slightly shallower depths; this unusual mantle region just above the core is called **D″** ("D double-prime"), a nomenclature introduced over 50 years ago by the geophysicist Keith Bullen. **D″** may consist of material from subducted slabs that descended and came to rest at the core–mantle boundary and/or from a new mineral polymorph discovered in perovskite called post-perovskite.

Earthquakes at shallow depths are a result of stick-slip faulting; however, below about 50 km (31 mi) the hot, high pressure conditions ought to inhibit further seismicity. The mantle is considered to be viscous and incapable of brittle faulting. However, in subduction zones, earthquakes are observed down to 670 km (420 mi). A number of mechanisms have been proposed to explain this phenomenon, including dehydration, thermal runaway, and phase change. The geothermal gradient can be lowered where cool material from the surface sinks downward, increasing the strength of the surrounding mantle, and allowing earthquakes to occur down to a depth of 400 km (250 mi) and 670 km (420 mi).

The pressure at the bottom of the mantle is ~136 GPa (1.4 million atm). Pressure increases as depth increases, since the material beneath has to support the weight of all the material above it. The entire mantle, however, is thought to deform like a fluid on long timescales, with permanent plastic deformation accommodated by the movement of point, line, and/or planar defects through the solid crystals comprising the mantle. Estimates for the viscosity of the upper mantle range between 10^{19} and 10^{24} Pa·s, depending on depth, temperature, composition, state of stress, and numerous other factors. Thus, the upper mantle can only flow very slowly. However, when large forces are applied to the uppermost mantle it can become weaker, and this effect is thought to be important in allowing the formation of tectonic plate boundaries.

Exploration

Exploration of the mantle is generally conducted at the seabed rather than on land because of the relative thinness of the oceanic crust as compared to the significantly thicker continental crust.

The first attempt at mantle exploration, known as Project Mohole, was abandoned in 1966 after repeated failures and cost over-runs. The deepest penetration was approximately 180 m (590 ft). In 2005 an oceanic borehole reached 1,416 metres (4,646 ft) below the sea floor from the ocean drilling vessel *JOIDES Resolution*.

On 5 March 2007, a team of scientists on board the RRS *James Cook* embarked on a voyage to an area of the Atlantic seafloor where the mantle lies exposed without any crust covering, midway between the Cape Verde Islands and the Caribbean Sea. The exposed site lies approximately three kilometres beneath the ocean surface and covers thousands of square kilometres. A relatively difficult attempt to retrieve samples from the Earth's mantle was scheduled for later in 2007. The Chikyu Hakken mission attempted to use the Japanese vessel *Chikyū* to drill up to 7,000 m (23,000 ft) below the seabed. This is nearly three times as deep as preceding oceanic drillings.

A novel method of exploring the uppermost few hundred kilometres of the Earth was recently proposed, consisting of a small, dense, heat-generating probe which melts its way down through the crust and mantle while its position and progress are tracked by acoustic signals generated in the rocks. The probe consists of an outer sphere of tungsten about one metre in diameter with a cobalt-60 interior acting as a radioactive heat source. It was calculated that such a probe will reach the oceanic Moho in less than 6 months and attain minimum depths of well over 100 km (62 mi) in a few decades beneath both oceanic and continental lithosphere.

Exploration can also be aided through computer simulations of the evolution of the mantle. In 2009, a supercomputer application provided new insight into the distribution of mineral deposits, especially isotopes of iron, from when the mantle developed 4.5 billion years ago.

Mantle Convection

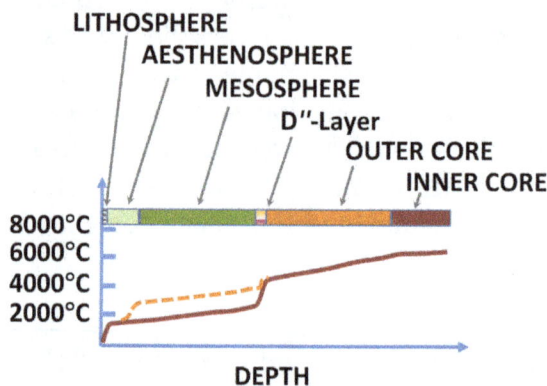

Calculated Earth's temperature vs. depth. Dashed curve: Layered mantle convection; Solid curve: Whole mantle convection.

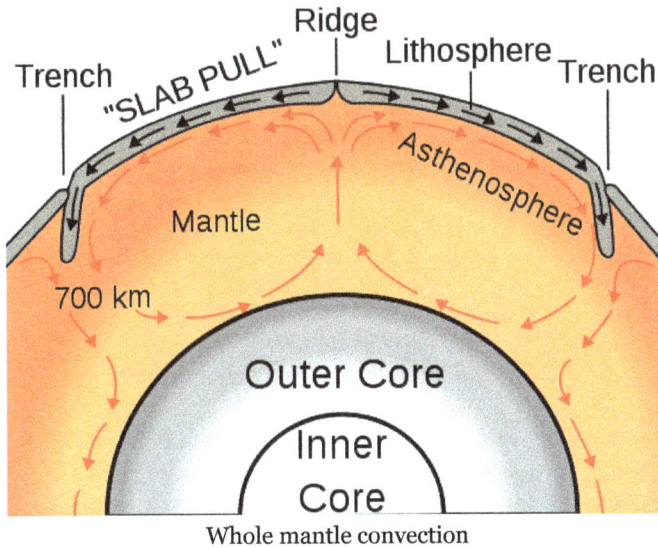
Whole mantle convection

Mantle convection is the slow creeping motion of Earth's solid silicate mantle caused by convection currents carrying heat from the interior of the Earth to the surface. The Earth's surface lithosphere, which rides atop the asthenosphere (the two components of the upper mantle), is divided into a number of plates that are continuously being created and consumed at their opposite plate boundaries. Accretion occurs as mantle is added to the growing edges of a plate, associated with seafloor spreading. This hot added material cools down by conduction and convection of heat. At the consumption edges of the plate, the material has thermally contracted to become dense, and it sinks under its own weight in the process of subduction usually at an ocean trench.

This *subducted* material sinks through the Earth's interior. Some subducted material appears to reach the lower mantle, while in other regions, this material is impeded from sinking further, possibly due to a phase transition from spinel to silicate perovskite and magnesiowustite, an endothermic reaction.

The subducted oceanic crust triggers volcanism, although the basic mechanisms are varied. Volcanism may occur due to processes that add buoyancy to partially melted mantle causing an upward flow due to a decrease in density of the partial melt.

Secondary forms of convection that may result in surface volcanism are postulated to occur as a consequence of intraplate extension and mantle plumes.

It is because the mantle can convect that the tectonic plates are able to move around the Earth's surface.

Mantle convection seems to have been much more active during the Hadean period, resulting in gravitational sorting of heavier molten iron, and nickel elements and sulphides in the core, and lighter silicate minerals in the mantle.

Types of Convection

During the late 20th century, there was significant debate within the geophysics community as to whether convection is likely to be 'layered' or 'whole'. Although elements of this debate still continue, results from seismic tomography, numerical simulations of mantle convection and examination of Earth's gravitational field are all beginning to suggest the existence of 'whole' mantle convection, at least at the present time. In this model, cold, subducting oceanic lithosphere descends all the way from the surface to the core-mantle boundary (CMB) and hot plumes rise from the CMB all the way to the surface. This picture is strongly based on the results of global seismic tomography models, which typically show slab and plume-like anomalies crossing the mantle transition zone.

Although it is now well accepted that subducting slabs cross the mantle transition zone and descend into the lower mantle, debate about the existence and continuity of plumes persists, with important implications for the style of mantle convection. This debate is linked to the controversy regarding whether intraplate volcanism is caused by shallow, upper-mantle processes or by plumes from the lower mantle. Many geochemistry studies have argued that the lavas erupted in intraplate areas are different in composition from shallow-derived mid ocean ridge basalts (MORB). Specifically, they typically have elevated Helium-3 - Helium-4 ratios. Being a primordial nuclide, Helium-3 is not naturally produced on earth. It also quickly escapes from earth's atmosphere when erupted. The elevated He-3/He-4 ratio of Ocean Island Basalts (OIBs) suggest that they must be sources from a part of the earth that has not previously been melted and reprocessed in the same way as MORB source has been. This has been interpreted as their originating from a different, less well-mixed, region, suggested to be the lower mantle. Others, however, have pointed out that geochemical differences could indicate the inclusion of a small component of near-surface material from the lithosphere.

Speed of Convection

Typical mantle convection speed is 20 mm/yr near the crust but can vary quite a bit. The small scale convection in the upper mantle is much faster than the convection near the core. A single shallow convection cycle takes on the order of 50 million years, though deeper convection can be closer to 200 million years.

Creep in the Mantle

Since the mantle is primarily composed of olivine ($(Mg,Fe)_2SiO_4$), the rheological characteristics of the mantle are largely those of olivine. Additionally, due to the varying temperatures and pressures between the lower and upper mantle, a variety of creep processes can occur with dislocation creep dominating in the lower mantle and diffusional creep occasionally dominating in the upper mantle. However, there is a large transition region in creep processes between the upper and lower mantle and even within each section, creep properties can change strongly with location and thus temperature and pressure. In the power law creep regions, the creep equation fitted to data with n = 3-4 is standard.

The strength of olivine not only scales with its melting temperature, but also is very sensitive to water and silica content. The solidus depression by impurities, primarily Ca, Al, and Na, and pressure affects creep behavior and thus contributes to the change in creep mechanisms with location. While creep behavior is generally plotted as homologous temperature versus stress, in the case of the mantle it is often more useful to look at the pressure dependence of stress. Though stress is simple force over area, defining the area is difficult in geology. Equation 1 demonstrates the pressure dependence of stress. Since it is very difficult to simulate the high pressures in the mantle (1MPa at 300–400 km), the low pressure laboratory data is usually extrapolated to high pressures by applying creep concepts from metallurgy.

$$(1) \quad \left(\frac{\partial \ln \sigma}{\partial P}\right)_{T,\dot{\epsilon}} = \left(\frac{1}{TT_m}\right) * \left(\frac{\partial \ln \sigma}{\partial (1/T)}\right)_{P,\dot{\epsilon}} * \frac{dT_m}{dP}$$

Most of the mantle has homologous temperatures of 0.65-0.75 and experiences strain rates of $10^{-14} - 10^{-16}$ 1/s. Stresses in mantle are dependent on density, gravity, thermal expansion coefficients, temperature differences driving convection, and distance convection occurs over, all of which give stresses around a fraction of 3-30MPa. Due to the large grain sizes (at low stresses as high as several mm), it is unlikely that Nabarro-Herring (NH) creep truly dominates. Given the large grain sizes, dislocation creep tends to dominate. 14 MPa is the stress below which diffusional creep dominates and above which power law creep dominates at 0.5Tm of olivine. Thus, even for relatively low temperatures, the stress diffusional creep would operate at is too low for realistic conditions. Though the power law creep rate increases with increasing water content due to weakening, reducing activation energy of diffusion and thus increasing the NH creep rate, NH is generally still not large enough to dominate. Nevertheless, diffusional creep can dominate in very cold or deep parts of the upper mantle. Additional deformation in the mantle can be attributed to transformation enhanced ductility. Below 400 km, the olivine undergoes a pressure induced phase transformation into spinel and can cause more deformation due to the increased ductility. Further evidence for the dominance of power law creep comes from preferred lattice orientations as a result of deformation. Under dislocation creep, crystal structures reorient into lower stress orientations. This does not happen under diffusional creep, thus observation of preferred orientations in samples lends credence to the dominance of dislocation creep.

Outer Core

The outer core of the Earth is a fluid layer about 2,300 km (1,400 mi) thick and composed of mostly iron and nickel that lies above Earth's solid inner core and below its mantle. Its outer boundary lies 2,890 km (1,800 mi) beneath Earth's surface. The transition between the inner core and outer core is located approximately 5,150 km (3,200 mi) beneath the Earth's surface. Unlike the inner core, the outer core is not solid. This is also referred to as the "liquid core".

Properties

Estimates for the temperature of the outer core are about 3,000–4,500 K (2,730–4,230 °C; 4,940–7,640 °F) in its outer regions and 4,000–8,000 K (3,730–7,730 °C; 6,740–13,940 °F) near the inner core. Evidence for a fluid outer core includes observations from seismology which shows that seismic shear-waves are not transmitted through the outer core. Because of its high temperature, modeling work has shown that the outer core is a low viscosity fluid that convects turbulently. Eddy currents in the nickel iron fluid of the outer core are believed to influence the Earth's magnetic field. The average magnetic field strength in the Earth's outer core was measured to be 2.5 millitesla, 50 times stronger than the magnetic field at the surface. The outer core is not under enough pressure to be solid, so it is liquid even though it has a composition similar to that of the inner core. Sulfur and oxygen could also be present in the outer core.

As heat is transferred outward toward the mantle, the net trend is for the inner boundary of the liquid region to freeze, causing the solid inner core to grow. This growth rate is estimated to be 1 mm per year.

Inner Core

The Earth's inner core is the Earth's innermost part and according to seismological studies, it has been believed to be primarily a solid ball with a radius of about 1,220 kilometres (760 miles) (about 70% of the Moon's radius). It is composed of an iron–nickel alloy and some light elements. The temperature at the inner core boundary is approximately 5700 K (5400 °C).

Discovery

The Earth was discovered to have a solid inner core distinct from its liquid outer core in 1936, by the Danish seismologist Inge Lehmann, who deduced its presence by studying seismograms from earthquakes in New Zealand. She observed that the seismic waves reflect off the boundary of the inner core and can be detected by sensitive seismographs on the Earth's surface. This boundary is known as the Bullen discontinuity, or sometimes as the Lehmann discontinuity. A few years later, in 1940, it was hypothesized that this inner core was made of solid iron; its rigidity was confirmed in 1971.

The outer core was determined to be liquid from observations showing that compressional waves pass through it, but elastic shear waves do not – or do so only very weakly.

The solidity of the inner core had been difficult to establish because the elastic shear waves that are expected to pass through a solid mass are very weak and difficult for seismographs on the Earth's surface to detect, since they become so attenuated on their way from the inner core to the surface by their passage through the liquid outer core. Dziewonski and Gilbert established that measurements of normal modes of vibration of Earth caused by large earthquakes were consistent with a liquid outer core. Recent claims that shear waves have been detected passing through the inner core were initially controversial, but are now gaining acceptance.

Composition

Based on the relative prevalence of various chemical elements in the Solar System, the theory of planetary formation, and constraints imposed or implied by the chemistry of the rest of the Earth's volume, the inner core is believed to consist primarily of a nickel-iron alloy. The iron-nickel alloy under core pressure is denser than the core, implying the presence of light elements in the core (e.g. silicon, oxygen, sulfur).

Temperature and Pressure

The temperature of the inner core can be estimated by considering both the theoretical and the experimentally demonstrated constraints on the melting temperature of impure iron at the pressure which iron is under at the boundary of the inner core (about 330 GPa). These considerations suggest that its temperature is about 5,700 K (5,400 °C; 9,800 °F). The pressure in the Earth's inner core is slightly higher than it is at the boundary between the outer and inner cores: it ranges from about 330 to 360 gigapascals (3,300,000 to 3,600,000 atm). Iron can be solid at such high temperatures only because its melting temperature increases dramatically at pressures of that magnitude.

A report published in the journal *Science* concludes that the melting temperature of iron at the inner core boundary is 6230 ± 500 K, roughly 1000 K higher than previous estimates.

Dynamics

The Earth's inner core is thought to be slowly growing as the liquid outer core at the boundary with the inner core cools and solidifies due to the gradual cooling of the Earth's interior (about 100 degrees Celsius per billion years). Many scientists had initially expected that the inner core would be found to be homogeneous, because the solid inner core was originally formed by a gradual cooling of molten material, and continues to grow as a result of that same process. Even though it is growing into liquid, it is solid, due to the very high pressure that keeps it compacted together even if the temperature is extremely high. It was even suggested that Earth's inner core might be a single crystal of iron. However, this prediction was disproved by observations indicating that in fact there is a degree of disorder within the inner core. Seismologists have found that the inner core is not completely uniform, but instead contains large-scale structures such that seismic waves pass more rapidly through some parts of the inner core than through others. In addition, the properties of the inner core's surface vary from place to place across distances as small as 1 km. This variation is surprising, since lateral temperature variations along the inner-core boundary are known to be extremely small (this conclusion is confidently constrained by magnetic field observations). Recent discoveries suggest that the solid inner core itself is composed of layers, separated by a transition zone about 250 to 400 km thick. If the inner core grows by small frozen sediments falling onto its surface, then some liquid can also be trapped in the pore spaces and some of this residual fluid may still persist to some small degree in much of its interior.

Because the inner core is not rigidly connected to the Earth's solid mantle, the possibility that it rotates slightly faster or slower than the rest of Earth has long been entertained. In the 1990s, seismologists made various claims about detecting this kind of super-rotation by observing changes in the characteristics of seismic waves passing through the inner core over several decades, using the aforementioned property that it transmits waves faster in some directions. Estimates of this super-rotation are around one degree of extra rotation per year.

Growth of the inner core is thought to play an important role in the generation of Earth's magnetic field by dynamo action in the liquid outer core. This occurs mostly because it cannot dissolve the same amount of light elements as the outer core and therefore freezing at the inner core boundary produces a residual liquid that contains more light elements than the overlying liquid. This causes it to become buoyant and helps drive convection of the outer core. The existence of the inner core also changes the dynamic motions of liquid in the outer core as it grows and may help fix the magnetic field since it is expected to be a great deal more resistant to flow than the outer core liquid (which is expected to be turbulent).

Speculation also continues that the inner core might have exhibited a variety of internal deformation patterns. This may be necessary to explain why seismic waves pass more rapidly in some directions than in others. Because thermal convection alone appears to be improbable, any buoyant convection motions will have to be driven by variations in composition or abundance of liquid in its interior. S. Yoshida and colleagues proposed a novel mechanism whereby deformation of the inner core can be caused by a higher rate of freezing at the equator than at polar latitudes, and S. Karato proposed that changes in the magnetic field might also deform the inner core slowly over time.

There is an East–West asymmetry in the inner core seismological data. There is a model which explains this by differences at the surface of the inner core – melting in one hemisphere and crystallization in the other. the western hemisphere of the inner core may be crystallizing, whereas the eastern hemisphere may be melting. This may lead to enhanced magnetic field generation in the crystallizing hemisphere, creating the asymmetry in the Earth's magnetic field.

History

Based on rates of cooling of the core, it is estimated that the current solid inner core started solidifying approximately 0.5 to 2 billion years ago out of a fully molten core (which formed just after planetary formation). If true, this would mean that the Earth's solid inner core is not a primordial feature that was present during the planet's formation, but a feature younger than the Earth (the Earth is about 4.5 billion years old).

Mesoplates

The concept of "mesoplates" was introduced as a heuristic for characterizing the motion of lithospheric plates relative to the sublithospheric source region of hotspot volcanism (Pilger, 2003). W. Jason Morgan (1972), originally suggested that hotspots (inferred by J. Tuzo Wilson) beneath such active volcanic regions as Hawaii and Iceland form a fixed "absolute" frame of reference for the motion of the overlying plates. However, the existence of a globally fixed reference frame for island-seamount chains and aseismic ridges ("traces") that are inferred to have originated from hotspots was quickly discounted by the primitive plate reconstructions available in the mid-1970s (Molnar and Atwater, 1973). Further, paleomagnetic measurements imply that hotspots have moved relative to the magnetic poles of the Earth (the magnetic poles are further inferred to correspond with the rotational poles of the planet when averaged over thousands of years). Aside: the term "hotspot" is used herein without any genetic implications. The term "melting spot" might well be more applicable.

Development of the Concept

As plate reconstructions have improved over the succeeding three decades since Morgan's original contribution, it is become apparent that the hotspots beneath the central North and South Atlantic and Indian Oceans may form one, distinct frame of reference, while those underlying the plates beneath the Pacific Ocean form a separate reference frame. For convenience, the hotspots beneath the Pacific Ocean are referred to as the "Hawaiian set" after Hawaii, while those beneath much of the Atlantic and Indian Ocean are called the "Tristan set" after the island of Tristan da Cunha (the Tristan hotspot), one of the principal inferred hotspots of the set. Within a single hotspot set, the traces tied to their originating hotspot can be fit by plate reconstructions which imply only minor relative motion among the hotspots for perhaps the past 130 m.y. (million years) for the Tristan set and 80 m.y. for the Hawaiian set. However, the two hotspot sets are inconsistent with the hypothesis of a single hotspot reference frame;

distinct motion between the two sets is apparent between 80 and 30 Ma (m.y. before Present; e.g., Raymond, et al., 2000).

It is important to acknowledge that radiometric dating of volcanism along hotspot traces may or may not accurately and precisely constrain the position of the plate above the underlying hotspot at the analytically produced age. However, reconstruction models for the Hawaiian set are constrained in age by the hotspot beneath Easter Island and its traces on the Pacific and Nazca plates between approximately 50 and 30 Ma, as the hotspot was beneath the spreading center during that time interval, and resulting relative plate reconstructions constrain motion of the plates relative to the hotspot. Prior to 50 Ma and since 30 Ma, reconstructions can be determined that fit virtually all existing Hawaiian set traces; the actual ages have the greatest uncertainty. Similarly, plate reconstructions relative to the Tristan set are best constrained in age by relative plate reconstructions, a fortuitous consequence of spherical plate tectonics of three or more plates.

Lithospheric plates are recognized in terms of their lack of internal deformation. Thus two points on the same plate will not move relative to one another, even if the plate moves relative to another plate (or relative to the Earth's rotational poles). Plates are not explicitly defined in terms of their mechanical properties. In a sense, then, "plates" are a heuristic—rather like fitting a straight line through a set of points without a clear functional relationship. Analogously, the term "mesoplate" was introduced. Since the hotspots of the Hawaiian set appear to form a frame of reference (like points on a lithospheric plate, they don't appear to be moving at a very great rate relative to one another), the hotspots and that part of the upper mantle in which they are embedded is termed the "Hawaiian mesoplate". The "Tristan mesoplate" is similarly defined. A third mesoplate, "Icelandic", is inferred to underlie the northernmost Atlantic Ocean, the Arctic Ocean, much of Eurasia to the north of the Alps and Himalayas; since the Iceland hotspot trace is not consistent with either the Hawaiian or Tristan set.

Additional evidence for mesoplates comes from observations that intraplate stresses in stable continental interiors of North America and Africa are consistent with plate motions in the Tristan hotspot frame. This observation was first made for contemporary stresses (the maximum horizontal principal compressive stress – sigma-hx); and also appears to hold for paleostress indicators between approximately 100 and 20 Ma (Pilger, 2003). This observation implies that the sublithospheric mantle over which the plates are moving comprises the same reference frame in which the hotspots are embedded.

The mesoplate heuristic is very much a hypothetical construct. Several observations could discount it. It is conceivable that a missing plate boundary between the plates beneath the Pacific and those beneath the Atlantic and Indian Oceans might be hidden and responsible for the discrepancy between the two hotspot sets. However, progressive study of the most likely region for such a boundary has failed to find it.

The origin of hotspots, whether from deep mantle plumes, mid-mantle melting anomalies, or intraplate fractures, is constrained somewhat by the mesoplate hypothesis. The principal alternative models for the origin of hotspot traces, propagating fractures, are still actively advocated by many workers. Such a model does not explicitly recognize sublithospheric reference frames. However, it cannot completely explain all of the features of the most familiar hotspot traces (Pilger, 2007).

The mantle plume hypothesis for the origin of hotspots need not be inconsistent with mesoplates. However, it would need to be modified to recognize that the lack of motion between hotspots represents a kind of "embedding" of the "plume" in the upper mantle (shallow mesosphere) of the Earth. One of Morgan's rationales for plumes was the existence of an "absolute motion" reference frame. Numerical modeling now indicates that such a reference frame would be unlikely in the context of plume convection.

If continued research were to demonstrate the continued applicability of the mesoplate hypothesis, it would have important implications for the nature of convection in the upper mantle: Convective motion beneath plates is almost entirely vertical within individual mesoplates; lateral motion in the mantle would be confined to mesoplate boundaries and to greater depths.

Origin of the Term

"Mesoplates" is a combination and contraction of two terms: "mesosphere", as applied to the solid earth, and "tectonic plates".

Mesosphere (Solid Earth)

"Mesosphere" is derived from "mesospheric shell", coined by Reginald Aldworth Daly, a Harvard University geology professor. In the pre-plate tectonics era, Daly (1940) inferred three spherical layers comprise the outer Earth: lithosphere (including the crust), asthenosphere, and mesospheric shell. Daly's hypothetical depths to the lithosphere–asthenosphere boundary ranged from 80 to 100 km and the top of the mesospheric shell (base of the asthenosphere) from 200 to 480 km. Thus, Daly's asthenosphere was inferred to be 120 to 400 km thick. According to Daly, the base of the solid earth mesosphere could extend to the base of the mantle (and, thus, to the top of the core).

Isacks, Oliver, and Sykes (1968) applied lithosphere and asthenosphere to their conception to the "New Global Tectonics" or what subsequently became known as plate tectonics. In their conception, the base of the asthenosphere extended as deep as the deepest (650–700 km) earthquakes in the inclined seismic zones where descending lithospheric plates penetrate the upper mantle.

Lithospheric (Tectonic) Plate

The (spherical) lithospheric plates of plate tectonics are so defined because they behave in a kinematically rigid manner. That is, any three points on the same plate do not move relative to one another, while the plate itself (and all points it contains) may move relative to other plates or other internal reference frames (e.g., the earth's spin axis or geomagnetic poles). In other words, ideal lithospheric plates do not deform internally as they move.

A "mesoplate", then behaves like lithospheric plates: empirical evidence (discussed above) indicates groups of melting anomalies (hotspots) embedded in the shallow mesosphere do not move relative to one another, but collectively move relative to other hotspot groups and relative to overlying lithospheric plates.

References

- Hart, P. J., Earth's Crust and Upper Mantle, American Geophysical Union, 1969, pp. 13-15 ISBN 978-0-87590-013-1

- Sorokhtin, O.G.; Chilingarian, G.V.; Sorokhtin, N.O. (2011). Evolution of Earth and its climate birth, life and death of Earth. Amsterdam: Elsevier Science Ltd. p. 137. ISBN 9780444537584. Retrieved 29 May 2015.

- Thompson, Graham R.; Turk, Jonathan (2007). Earth science and the environment (4th ed., International student edition. ed.). Australia: Thomson Brooks/Cole. pp. 133–134. ISBN 9780495112877. Retrieved 29 May 2015.

- Burns, Roger George (1993). Mineralogical Applications of Crystal Field Theory. Cambridge University Press. p. 354. ISBN 0-521-43077-1. Retrieved 2007-12-26.

- Anderson, Don L. (2007) New Theory of the Earth. Cambridge University Press. ISBN 978-0-521-84959-3, ISBN 0-521-84959-4

- Jackson, Ian (1998). The Earth's Mantle - Composition, Structure, and Evolution. Cambridge University Press. pp. 311–378. ISBN 0-521-78566-9.

- Kent C. Condie (1997). Plate tectonics and crustal evolution (4th ed.). Butterworth-Heinemann. p. 5. ISBN 0-7506-3386-7.

- Ctirad Matyska & David A Yuen (2007). "Figure 17 in Lower-mantle material properties and convection models of multiscale plumes". Plates, plumes, and planetary processes. Geological Society of America. p. 159. ISBN 0-8137-2430-9.

- Gerald Schubert; Donald Lawson Turcotte; Peter Olson (2001). "Chapter 2: Plate tectonics". Mantle convection in the earth and planets. Cambridge University Press. p. 16 ff. ISBN 0-521-79836-1.

- Gerald Schubert; Donald Lawson Turcotte; Peter Olson. "§2.5.3: Fate of descending slabs". Cited work. p. 35 ff. ISBN 0-521-79836-1.

- Foulger, G.R. (2010). Plates vs. Plumes: A Geological Controversy. Wiley-Blackwell. ISBN 978-1-4051-6148-0.

- Donald Lawson Turcotte; Gerald Schubert (2002). Geodynamics (2nd ed.). Cambridge University Press. ISBN 0-521-66624-4.

- Hung Kan Lee (2002). International handbook of earthquake and engineering seismology; volume 1. Academic Press. p. 926. ISBN 0-12-440652-1.

- Rudoph, Maxwell (11 December 2015). "Viscosity jump in Earth's mid-mantle". Science. Retrieved 16 January 2016.

- Sumner, Thomas (10 December 2015). "Gooey rock in mantle thickens 1,000 kilometers down". Science News. Retrieved 16 January 2016.

- Garnero, Edward J.; McNamara, Allen K. and Shim, Sang-Heon (2016). "Continent-sized anomalous zones with low seismic velocity at the base of Earth's mantle". Nature Geoscience. 9: 481–489. doi:10.1038/NGEO2733.

- Longhi, John; et al. (1992). "The bulk composition, mineralogy and internal structure of Mars". Mars (A93-27852 09-91). University of Arizona Press, Tucson. pp. 184–208. Retrieved 16 October 2015.

- "MESSENGER Provides New Look at Mercury's surprising core and landscape curiosities". NASA. 21 March 2012. Retrieved 16 October 2015.

- NASA (6 October 2000). "Scientists Show Jovian Moon Io's Mantle is Similar to Earth". NASA. Retrieved 7 October 2015.

- "Rare Diamond confirms that Earth's mantle holds an ocean's worth of water". Scientific American. March 12, 2014. Retrieved March 13, 2014.

- Neumann, W.; et al. (2014). "Differentiation of Vesta: Implications for a shallow magma ocean". Earth and Planetary Science Letters. 395: 267–280. doi:10.1016/j.epsl.2014.03.033.

- Breaking News | Oldest rock shows Earth was a hospitable young planet. Spaceflight Now (2001-01-14). Retrieved on 2012-01-27.

- Lawrence Berkeley National Laboratory (Berkeley Lab) is a Department of Energy (DOE) Office of Science lab managed by University of California., What Keeps the Earth Cooking? News Release by Paul Preuss, July 17, 2011

- Ozawa, H.; al., et (2011). "Phase Transition of FeO and Stratification in Earth's Outer Core". Science. 334 (6057): 792–794. Bibcode:2011Sci...334..792O. doi:10.1126/science.1208265.

- Lawrence Berkeley National Laboratory (Berkeley Lab) is a Department of Energy (DOE) Office of Science lab managed by University of California., What Keeps the Earth Cooking? News Release by Paul Preuss, July 17, 2011

- Gubbins, David; Sreenivasan, Binod; Mound, Jon; Rost, Sebastian (May 19, 2011). "Melting of the Earth's inner core". Nature. 473 (7347): 361–363. Bibcodc:2011Natur.473..361G.

- doi:10.1038/nature10068. PMID 21593868.

- Waszek, Lauren; Irving, Jessica; Deuss, Arwen (2011). "Reconciling the hemispherical structure of Earth's inner core with its super-rotation". Nature Geoscience. 4: 264–267. doi:10.1038/ngeo1083.

- John C. Butler (1995). "Class Notes - The Earth's Interior". Physical Geology Grade Book. University of Houston. Retrieved 30 August 2011.

5

Earthquake: A Comprehensive Study

An earthquake is the shaking of the surface of the Earth. Earthquakes can severely damage life and property and can also result in decline in the economy. It can cause damage that can take years to recover from, and they can be measured by using observations from seismometers. Some of the aspects elucidated within this section are aftershock, foreshock, induced seismicity, cryoseism and submarine earthquake.

Earthquake

An earthquake (also known as a quake, tremor or temblor) is the perceptible shaking of the surface of the Earth, resulting from the sudden release of energy in the Earth's crust that creates seismic waves. Earthquakes can be violent enough to toss people around and destroy whole cities. The seismicity or seismic activity of an area refers to the frequency, type and size of earthquakes experienced over a period of time.

Preliminary Determination of Epicenters
358,214 Events, 1963 - 1998

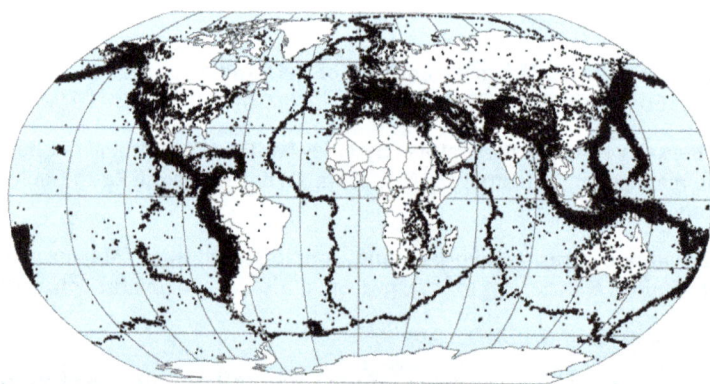

Global earthquake epicenters, 1963–1998

Earthquakes are measured using observations from seismometers. The moment magnitude is the most common scale on which earthquakes larger than approximately 5 are reported for the entire globe. The more numerous earthquakes smaller than magnitude 5 reported by national seismological observatories are measured mostly on the local magnitude scale, also referred to as the Richter magnitude scale. These two scales are numerically similar over their range of validity. Magnitude 3 or lower earthquakes are mostly imperceptible or weak and magnitude 7 and over potentially cause serious damage over larger areas, depending on their depth. The largest earthquakes in historic times have been of magnitude slightly over 9, although there is no limit to the possible magnitude. Intensity of shaking is measured on the modified Mercalli scale. The shallower an earthquake, the more damage to structures it causes, all else being equal.

At the Earth's surface, earthquakes manifest themselves by shaking and sometimes displacement of the ground. When the epicenter of a large earthquake is located offshore, the seabed may be displaced sufficiently to cause a tsunami. Earthquakes can also trigger landslides, and occasionally volcanic activity.

In its most general sense, the word *earthquake* is used to describe any seismic event — whether natural or caused by humans — that generates seismic waves. Earthquakes are caused mostly by rupture of geological faults, but also by other events such as volcanic activity, landslides, mine blasts, and nuclear tests. An earthquake's point of initial rupture is called its focus or hypocenter. The epicenter is the point at ground level directly above the hypocenter.

Naturally Occurring Earthquakes

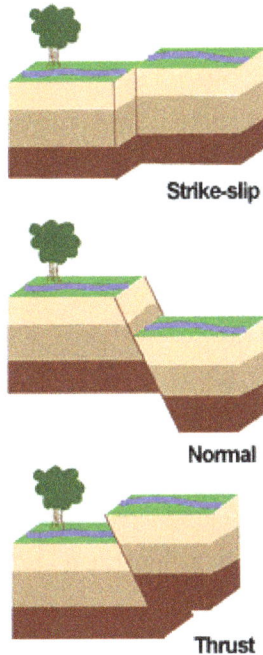

Fault types

Tectonic earthquakes occur anywhere in the earth where there is sufficient stored elastic strain energy to drive fracture propagation along a fault plane. The sides of a fault move past each other smoothly and aseismically only if there are no irregularities or asperities along the fault surface that increase the frictional resistance. Most fault surfaces do have such asperities and this leads to a form of stick-slip behavior. Once the fault has locked, continued relative motion between the plates leads to increasing stress and therefore, stored strain energy in the volume around the fault surface. This continues until the stress has risen sufficiently to break through the asperity, suddenly allowing sliding over the locked portion of the fault, releasing the stored energy. This energy is released as a combination of radiated elastic strain seismic waves, frictional heating of the fault surface, and cracking of the rock, thus causing an earthquake. This process of gradual build-up of strain and stress punctuated by occasional sudden earthquake failure is referred to as the elastic-rebound theory. It is estimated that only 10 percent or less of an earthquake's total energy is radiated as seismic energy. Most of the earthquake's energy is used to power the earthquake fracture growth or is converted into heat generated by friction. Therefore, earthquakes lower the Earth's available elastic potential energy and raise its temperature, though these changes are negligible compared to the conductive and convective flow of heat out from the Earth's deep interior.

Earthquake Fault Types

There are three main types of fault, all of which may cause an interplate earthquake: normal, reverse (thrust) and strike-slip. Normal and reverse faulting are examples of dip-slip, where the displacement along the fault is in the direction of dip and movement on them involves a vertical component. Normal faults occur mainly in areas where the crust is being extended such as a divergent boundary. Reverse faults occur in areas where the crust is being shortened such as at a convergent boundary. Strike-slip faults are steep structures where the two sides of the fault slip horizontally past each other; transform boundaries are a particular type of strike-slip fault. Many earthquakes are caused by movement on faults that have components of both dip-slip and strike-slip; this is known as oblique slip.

Reverse faults, particularly those along convergent plate boundaries are associated with the most powerful earthquakes, megathrust earthquakes, including almost all of those of magnitude 8 or more. Strike-slip faults, particularly continental transforms, can produce major earthquakes up to about magnitude 8. Earthquakes associated with normal faults are generally less than magnitude 7. For every unit increase in magnitude, there is a roughly thirtyfold increase in the energy released. For instance, an earthquake of magnitude 6.0 releases approximately 30 times more energy than a 5.0 magnitude earthquake and a 7.0 magnitude earthquake releases 900 times (30 × 30) more energy than a 5.0 magnitude of earthquake. An 8.6 magnitude earthquake releases the same amount of energy as 10,000 atomic bombs like those used in World War II.

This is so because the energy released in an earthquake, and thus its magnitude, is proportional to the area of the fault that ruptures and the stress drop. Therefore, the longer the length and the wider the width of the faulted area, the larger the resulting magnitude. The topmost, brittle part of the Earth's crust, and the cool slabs of the tectonic plates that are descending down into the hot mantle, are the only parts of our planet which can store elastic energy and release it in fault ruptures. Rocks hotter than about 300 degrees Celsius flow in response to stress; they do not rupture in earthquakes. The maximum observed lengths of ruptures and mapped faults (which may break in a single rupture) are approximately 1000 km. Examples are the earthquakes in Chile, 1960; Alaska, 1957; Sumatra, 2004, all in subduction zones. The longest earthquake ruptures on strike-slip faults, like the San Andreas Fault (1857, 1906), the North Anatolian Fault in Turkey (1939) and the Denali Fault in Alaska (2002), are about half to one third as long as the lengths along subducting plate margins, and those along normal faults are even shorter.

Aerial photo of the San Andreas Fault in the Carrizo Plain, northwest of Los Angeles

The most important parameter controlling the maximum earthquake magnitude on a fault is however not the maximum available length, but the available width because the latter varies by a factor of 20. Along converging plate margins, the dip angle of the rupture plane is very shallow, typically about 10 degrees. Thus the width of the plane within the top brittle crust of the Earth can become 50 to 100 km (Japan, 2011; Alaska, 1964), making the most powerful earthquakes possible.

Strike-slip faults tend to be oriented near vertically, resulting in an approximate width of 10 km within the brittle crust, thus earthquakes with magnitudes much larger than 8 are not possible. Maximum magnitudes along many normal faults are even more limited because many of them are located along spreading centers, as in Iceland, where the thickness of the brittle layer is only about 6 km.

In addition, there exists a hierarchy of stress level in the three fault types. Thrust faults are generated by the highest, strike slip by intermediate, and normal faults by the lowest stress levels. This can easily be understood by considering the direction of the greatest principal stress, the direction of the force that 'pushes' the rock mass during the faulting. In the case of normal faults, the rock mass is pushed down in a vertical direction, thus the pushing force (greatest principal stress) equals the weight of the rock mass itself. In the case of thrusting, the rock mass 'escapes' in the direction of the least principal stress, namely upward, lifting the rock mass up, thus the overburden equals the **least** principal stress. Strike-slip faulting is intermediate between the other two types described above. This difference in stress regime in the three faulting environments can contribute to differences in stress drop during faulting, which contributes to differences in the radiated energy, regardless of fault dimensions.

Earthquakes away from Plate Boundaries

Where plate boundaries occur within the continental lithosphere, deformation is spread out over a much larger area than the plate boundary itself. In the case of the San Andreas fault continental transform, many earthquakes occur away from the plate boundary and are related to strains developed within the broader zone of deformation caused by major irregularities in the fault trace (e.g., the "Big bend" region). The Northridge earthquake was associated with movement on a blind thrust within such a zone. Another example is the strongly oblique convergent plate boundary between the Arabian and Eurasian plates where it runs through the northwestern part of the Zagros Mountains. The deformation associated with this plate boundary is partitioned into nearly pure thrust sense movements perpendicular to the boundary over a wide zone to the southwest and nearly pure strike-slip motion along the Main Recent Fault close to the actual plate boundary itself. This is demonstrated by earthquake focal mechanisms.

All tectonic plates have internal stress fields caused by their interactions with neighboring plates and sedimentary loading or unloading (e.g. deglaciation). These stresses may be sufficient to cause failure along existing fault planes, giving rise to intraplate earthquakes.

Shallow-focus and Deep-focus Earthquakes

Collapsed Gran Hotel building in the San Salvador metropolis,
after the shallow 1986 San Salvador earthquake.

The majority of tectonic earthquakes originate at the ring of fire in depths not exceeding tens of kilometers. Earthquakes occurring at a depth of less than 70 km are classified as 'shallow-focus' earthquakes, while those with a focal-depth between 70 and 300 km are commonly termed 'mid-focus' or 'intermediate-depth' earthquakes. In subduction zones, where older and colder oceanic crust descends beneath another tectonic plate, Deep-focus earthquakes may occur at much greater depths (ranging from 300 up to 700 kilometers). These seismically active areas of subduction are known as Wadati–Benioff zones. Deep-focus earthquakes occur at a depth where the subducted lithosphere should no longer be brittle, due to the high temperature and pressure. A possible mechanism for the generation of deep-focus earthquakes is faulting caused by olivine undergoing a phase transition into a spinel structure.

Earthquakes and Volcanic Activity

Earthquakes often occur in volcanic regions and are caused there, both by tectonic faults and the movement of magma in volcanoes. Such earthquakes can serve as an early warning of volcanic eruptions, as during the 1980 eruption of Mount St. Helens. Earthquake swarms can serve as markers for the location of the flowing magma throughout the volcanoes. These swarms can be recorded by seismometers and tiltmeters (a device that measures ground slope) and used as sensors to predict imminent or upcoming eruptions.

Rupture Dynamics

A tectonic earthquake begins by an initial rupture at a point on the fault surface, a process known as nucleation. The scale of the nucleation zone is uncertain, with some evidence, such as the rupture dimensions of the smallest earthquakes, suggesting that it is smaller than 100 m while other evidence, such as a slow component revealed by low-frequency spectra of some earthquakes, suggest that it is larger. The possibility that the nucleation involves some sort of preparation process is supported by the observation that about 40% of earthquakes are preceded by foreshocks. Once the rupture has initiated, it begins to propagate along the fault surface. The mechanics of this process are poorly understood, partly because it is difficult to recreate the high sliding velocities in a laboratory. Also the effects of strong ground motion make it very difficult to record information close to a nucleation zone.

Rupture propagation is generally modeled using a fracture mechanics approach, likening the rupture to a propagating mixed mode shear crack. The rupture velocity is a function of the fracture energy in the volume around the crack tip, increasing with decreasing fracture energy. The velocity of rupture propagation is orders of magnitude faster than the displacement velocity across the fault. Earthquake ruptures typically propagate at velocities that are in the range 70–90% of the S-wave velocity, and this is independent of earthquake size. A small subset of earthquake ruptures appear to have propagated at speeds greater than the S-wave velocity. These supershear earthquakes have all been observed during large strike-slip events. The unusually wide zone of coseismic damage caused by the 2001 Kunlun earthquake has been attributed to the effects of the sonic boom developed in such earthquakes. Some earthquake ruptures travel at unusually low velocities and are referred to as slow earthquakes. A particularly dangerous form of slow earthquake is the tsunami earthquake, observed where the relatively low felt intensities, caused by the slow propagation speed of some great earthquakes, fail to alert the population of the neighboring coast, as in the 1896 Sanriku earthquake.

Tidal Forces

Tides may induce some seismicity.

Earthquake Clusters

Most earthquakes form part of a sequence, related to each other in terms of location and time. Most earthquake clusters consist of small tremors that cause little to no damage, but there is a theory that earthquakes can recur in a regular pattern.

Aftershocks

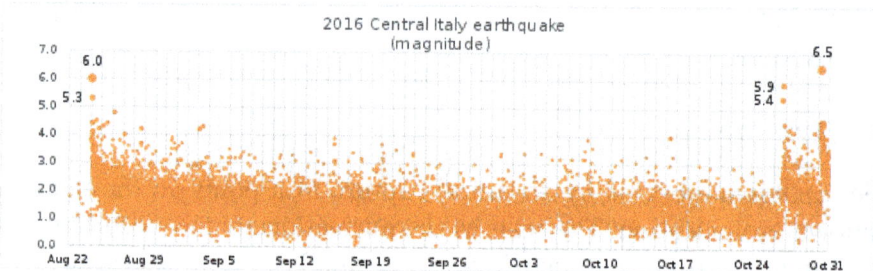

Magnitude of the Central Italy earthquakes of August and October 2016 and the aftershocks (which continued to occur after the period shown here).

An aftershock is an earthquake that occurs after a previous earthquake, the mainshock. An aftershock is in the same region of the main shock but always of a smaller magnitude. If an aftershock is larger than the main shock, the aftershock is redesignated as the main shock and the original main shock is redesignated as a foreshock. Aftershocks are formed as the crust around the displaced fault plane adjusts to the effects of the main shock.

Earthquake Swarms

Earthquake swarms are sequences of earthquakes striking in a specific area within a short period of time. They are different from earthquakes followed by a series of aftershocks by the fact that no single earthquake in the sequence is obviously the main shock, therefore none have notable higher magnitudes than the other. An example of an earthquake swarm is the 2004 activity at Yellowstone National Park. In August 2012, a swarm of earthquakes shook Southern California's Imperial Valley, showing the most recorded activity in the area since the 1970s.

Sometimes a series of earthquakes occur in what has been called an *earthquake storm*, where the earthquakes strike a fault in clusters, each triggered by the shaking or stress redistribution of the previous earthquakes. Similar to aftershocks but on adjacent segments of fault, these storms occur over the course of years, and with some of the later earthquakes as damaging as the early ones. Such a pattern was observed in the sequence of about a dozen earthquakes that struck the North Anatolian Fault in Turkey in the 20th century and has been inferred for older anomalous clusters of large earthquakes in the Middle East.

Size and Frequency of Occurrence

It is estimated that around 500,000 earthquakes occur each year, detectable with current instrumentation. About 100,000 of these can be felt. Minor earthquakes occur nearly constantly around the world in places like California and Alaska in the U.S., as well as in El Salvador, Mexico, Guatemala, Chile, Peru, Indonesia, Iran, Pakistan, the Azores in Portugal, Turkey, New Zealand, Greece, Italy, India, Nepal and Japan, but earthquakes can occur almost anywhere, including Downstate New York, England, and Australia. Larger earthquakes occur less frequently, the relationship being exponential; for example, roughly ten times as many earthquakes larger than magnitude 4 occur in a particular time period than earthquakes larger than magnitude 5. In the (low seismicity) United Kingdom, for example, it has been calculated that the average recurrences are: an earthquake of 3.7–4.6 every year, an earthquake of 4.7–5.5 every 10 years, and an earthquake of 5.6 or larger every 100 years. This is an example of the Gutenberg–Richter law.

The Messina earthquake and tsunami took as many as 200,000 lives on December 28, 1908 in Sicily and Calabria.

The number of seismic stations has increased from about 350 in 1931 to many thousands today. As a result, many more earthquakes are reported than in the past, but this is because of the vast improvement in instrumentation, rather than an increase in the number of earthquakes. The United States Geological Survey estimates that, since 1900, there have been an average of 18 major earthquakes (magnitude 7.0–7.9) and one great earthquake (magnitude 8.0 or greater) per year, and that this average has been relatively stable. In recent years, the number of major earthquakes per year has decreased, though this is probably a statistical fluctuation rather than a systematic trend. More detailed statistics on the size and frequency of earthquakes is available from the United States Geological Survey (USGS). A recent increase in the number of major earthquakes has been noted, which could be explained by a cyclical pattern of periods of intense tectonic activity, interspersed with longer periods of low-intensity. However, accurate recordings of earthquakes only began in the early 1900s, so it is too early to categorically state that this is the case.

Most of the world's earthquakes (90%, and 81% of the largest) take place in the 40,000 km long, horseshoe-shaped zone called the circum-Pacific seismic belt, known as the Pacific Ring of Fire, which for the most part bounds the Pacific Plate. Massive earthquakes tend to occur along other plate boundaries, too, such as along the Himalayan Mountains.

With the rapid growth of mega-cities such as Mexico City, Tokyo and Tehran, in areas of high seismic risk, some seismologists are warning that a single quake may claim the lives of up to 3 million people.

Induced Seismicity

While most earthquakes are caused by movement of the Earth's tectonic plates, human activity can also produce earthquakes. Four main activities contribute to this phenomenon: storing large amounts of water behind a dam (and possibly building an extremely heavy building), drilling and injecting liquid into wells, and by coal mining and oil drilling. Perhaps the best known example is the 2008 Sichuan earthquake in China's Sichuan Province in May; this tremor resulted in 69,227 fatalities and is the 19th deadliest earthquake of all time. The Zipingpu Dam is believed to have fluctuated the pressure of the fault 1,650 feet (503 m) away; this pressure probably increased the power of the earthquake and accelerated the rate of movement for the fault. The greatest earthquake in Australia's history is also claimed to be induced by humanity, through coal mining. The city of Newcastle was built over a large sector of coal mining areas. The earthquake has been reported to be spawned from a fault that reactivated due to the millions of tonnes of rock removed in the mining process.

Measuring and Locating Earthquakes

Earthquakes can be recorded by seismometers up to great distances, because seismic waves travel through the whole Earth's interior. The absolute magnitude of a quake is conventionally reported by numbers on the moment magnitude scale (formerly Richter scale, magnitude 7 causing serious damage over large areas), whereas the felt magnitude is reported using the modified Mercalli intensity scale (intensity II–XII).

Every tremor produces different types of seismic waves, which travel through rock with different velocities:

- Longitudinal P-waves (shock- or pressure waves)
- Transverse S-waves (both body waves)
- Surface waves — (Rayleigh and Love waves)

Propagation velocity of the seismic waves ranges from approx. 3 km/s up to 13 km/s, depending on the density and elasticity of the medium. In the Earth's interior the shock- or P waves travel much faster than the S waves (approx. relation 1.7 : 1). The differences in travel time from the epicenter to the observatory are a measure of the distance and can be used to image both sources of quakes and structures within the Earth. Also the depth of the hypocenter can be computed roughly.

In solid rock P-waves travel at about 6 to 7 km per second; the velocity increases within the deep mantle to ~13 km/s. The velocity of S-waves ranges from 2–3 km/s in light sediments and 4–5 km/s in the Earth's crust up to 7 km/s in the deep mantle. As a consequence, the first waves of a distant earthquake arrive at an observatory via the Earth's mantle.

On average, the kilometer distance to the earthquake is the number of seconds between the P and S wave times 8. Slight deviations are caused by inhomogeneities of subsurface structure. By such analyses of seismograms the Earth's core was located in 1913 by Beno Gutenberg.

Earthquakes are not only categorized by their magnitude but also by the place where they occur. The world is divided into 754 Flinn–Engdahl regions (F-E regions), which are based on political and geographical boundaries as well as seismic activity. More active zones are divided into smaller F-E regions whereas less active zones belong to larger F-E regions.

Standard reporting of earthquakes includes its magnitude, date and time of occurrence, geographic coordinates of its epicenter, depth of the epicenter, geographical region, distances to population centers, location uncertainty, a number of parameters that are included in USGS earthquake reports (number of stations reporting, number of observations, etc.), and a unique event ID.

Effects of Earthquakes

The effects of earthquakes include, but are not limited to, the following:

Shaking and Ground Rupture

Shaking and ground rupture are the main effects created by earthquakes, principally resulting in more or less severe damage to buildings and other rigid structures. The severity of the local effects depends on the complex combination of the earthquake magnitude, the distance from the epicenter, and the local geological and geomorphological conditions, which may amplify or reduce wave propagation. The ground-shaking is measured by ground acceleration.

Damaged buildings in Port-au-Prince, Haiti, January 2010.

1755 copper engraving depicting Lisbon in ruins and in flames after the 1755 Lisbon earthquake, which killed an estimated 60,000 people. A tsunami overwhelms the ships in the harbor.

Specific local geological, geomorphological, and geostructural features can induce high levels of shaking on the ground surface even from low-intensity earthquakes. This effect is called site or local amplification. It is principally due to the transfer of the seismic motion from hard deep soils to soft superficial soils and to effects of seismic energy focalization owing to typical geometrical setting of the deposits.

Ground rupture is a visible breaking and displacement of the Earth's surface along the trace of the fault, which may be of the order of several meters in the case of major earthquakes. Ground rupture is a major risk for large engineering structures such as dams, bridges and nuclear power stations and requires careful mapping of existing faults to identify any which are likely to break the ground surface within the life of the structure.

Landslides and Avalanches

Earthquakes, along with severe storms, volcanic activity, coastal wave attack, and wildfires, can produce slope instability leading to landslides, a major geological hazard. Landslide danger may persist while emergency personnel are attempting rescue.

Fires

Earthquakes can cause fires by damaging electrical power or gas lines. In the event of water mains rupturing and a loss of pressure, it may also become difficult to stop the spread of a fire once it has started. For example, more deaths in the 1906 San Francisco earthquake were caused by fire than by the earthquake itself.

Fires of the 1906 San Francisco earthquake

Soil Liquefaction

Soil liquefaction occurs when, because of the shaking, water-saturated granular material (such as sand) temporarily loses its strength and transforms from a solid to a liquid. Soil liquefaction may cause rigid structures, like buildings and bridges, to tilt or sink into the liquefied deposits. For example, in the 1964 Alaska earthquake, soil liquefaction caused many buildings to sink into the ground, eventually collapsing upon themselves.

Tsunami

The tsunami of the 2004 Indian Ocean earthquake

Tsunamis are long-wavelength, long-period sea waves produced by the sudden or abrupt movement of large volumes of water. In the open ocean the distance between wave crests can surpass 100 kilometers (62 mi), and the wave periods can vary from five minutes to one hour. Such tsunamis travel 600-800 kilometers per hour (373–497 miles per hour), depending on water depth. Large waves produced by an earthquake or a submarine landslide can overrun nearby coastal areas in a matter of minutes. Tsunamis can also travel thousands of kilometers across open ocean and wreak destruction on far shores hours after the earthquake that generated them.

Ordinarily, subduction earthquakes under magnitude 7.5 on the Richter scale do not cause tsunamis, although some instances of this have been recorded. Most destructive tsunamis are caused by earthquakes of magnitude 7.5 or more.

Floods

A flood is an overflow of any amount of water that reaches land. Floods occur usually when the volume of water within a body of water, such as a river or lake, exceeds the total capacity of the formation, and as a result some of the water flows or sits outside of the normal perimeter of the body. However, floods may be secondary effects of earthquakes, if dams are damaged. Earthquakes may cause landslips to dam rivers, which collapse and cause floods.

The terrain below the Sarez Lake in Tajikistan is in danger of catastrophic flood if the landslide dam formed by the earthquake, known as the Usoi Dam, were to fail during a future earthquake. Impact projections suggest the flood could affect roughly 5 million people.

Human Impacts

Ruins of the Għajn Ħadid Tower, which collapsed in an earthquake in 1856

An earthquake may cause injury and loss of life, road and bridge damage, general property damage, and collapse or destabilization (potentially leading to future collapse) of buildings. The aftermath may bring disease, lack of basic necessities, mental consequences such as panic attacks, depression to survivors, and higher insurance premiums.

Major Earthquakes

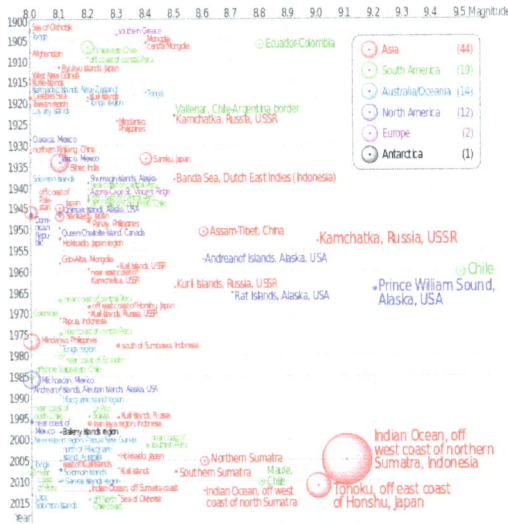

Asia (44)
South America (13)
Australia/Oceania (14)
North America (12)
Europe (2)
Antarctica (1)

Earthquakes of magnitude 8.0 and greater since 1900. The apparent 3D volumes of the bubbles are linearly proportional to their respective fatalities.

One of the most devastating earthquakes in recorded history was the 1556 Shaanxi earthquake, which occurred on 23 January 1556 in Shaanxi province, China. More than 830,000 people died. Most houses in the area were yaodongs—dwellings carved out of loess hillsides—and many victims were killed when these structures collapsed. The 1976 Tangshan earthquake, which killed between 240,000 and 655,000 people, was the deadliest of the 20th century.

The 1960 Chilean earthquake is the largest earthquake that has been measured on a seismograph, reaching 9.5 magnitude on 22 May 1960. Its epicenter was near Cañete, Chile. The energy released was approximately twice that of the next most powerful earthquake, the Good Friday earthquake (March 27, 1964) which was centered in Prince William Sound, Alaska. The ten largest recorded earthquakes have all been megathrust earthquakes; however, of these ten, only the 2004 Indian Ocean earthquake is simultaneously one of the deadliest earthquakes in history.

Earthquakes that caused the greatest loss of life, while powerful, were deadly because of their proximity to either heavily populated areas or the ocean, where earthquakes often create tsunamis that can devastate communities thousands of kilometers away. Regions most at risk for great loss of life include those where earthquakes are relatively rare but powerful, and poor regions with lax, unenforced, or nonexistent seismic building codes.

Prediction

Many methods have been developed for predicting the time and place in which earthquakes will occur. Despite considerable research efforts by seismologists, scientifically reproducible predictions cannot yet be made to a specific day or month. However, for well-understood faults the probability that a segment may rupture during the next few decades can be estimated.

Earthquake warning systems have been developed that can provide regional notification of an earthquake in progress, but before the ground surface has begun to move, potentially allowing people within the system's range to seek shelter before the earthquake's impact is felt.

Preparedness

The objective of earthquake engineering is to foresee the impact of earthquakes on buildings and other structures and to design such structures to minimize the risk of damage. Existing structures can be modified by seismic retrofitting to improve their resistance to earthquakes. Earthquake insurance can provide building owners with financial protection against losses resulting from earthquakes.

Emergency management strategies can be employed by a government or organization to mitigate risks and prepare for consequences.

Historical Views

Tremblement de terre en Italie. 340 ans avant J.-C. — L. Papirius
Cursor consul (d'après Lycosthène).

An image from a 1557 book

From the lifetime of the Greek philosopher Anaxagoras in the 5th century BCE to the 14th century CE, earthquakes were usually attributed to "air (vapors) in the cavities of the Earth." Thales of Miletus, who lived from 625–547 (BCE) was the only documented person who believed that earthquakes were caused by tension between the earth and water. Other theories existed, including the Greek philosopher Anaxamines' (585–526 BCE) beliefs that short incline episodes of dryness and wetness caused seismic activity. The Greek philosopher Democritus (460–371 BCE) blamed water in general for earthquakes. Pliny the Elder called earthquakes "underground thunderstorms."

Recent Studies

In recent studies, geologists claim that global warming is one of the reasons for increased seismic activity. According to these studies melting glaciers and rising sea levels disturb the balance of pressure on Earth's tectonic plates thus causing increase in the frequency and intensity of earthquakes.

Earthquakes in Culture

Mythology and Religion

In Norse mythology, earthquakes were explained as the violent struggling of the god Loki. When Loki, god of mischief and strife, murdered Baldr, god of beauty and light, he was punished by being bound in a cave with a poisonous serpent placed above his head dripping venom. Loki's wife Sigyn stood by him with a bowl to catch the poison, but whenever she had to empty the bowl the poison dripped on Loki's face, forcing him to jerk his head away and thrash against his bonds, which caused the earth to tremble.

In Greek mythology, Poseidon was the cause and god of earthquakes. When he was in a bad mood, he struck the ground with a trident, causing earthquakes and other calamities. He also used earthquakes to punish and inflict fear upon people as revenge.

In Japanese mythology, Namazu is a giant catfish who causes earthquakes. Namazu lives in the mud beneath the earth, and is guarded by the god Kashima who restrains the fish with a stone. When Kashima lets his guard fall, Namazu thrashes about, causing violent earthquakes.

In Popular Culture

In modern popular culture, the portrayal of earthquakes is shaped by the memory of great cities laid waste, such as Kobe in 1995 or San Francisco in 1906. Fictional earthquakes tend to strike suddenly and without warning. For this reason, stories about earthquakes generally begin with the disaster and focus on its immediate aftermath, as in *Short Walk to Daylight* (1972), *The Ragged Edge* (1968) or *Aftershock: Earthquake in New York* (1999). A notable example is Heinrich von Kleist's classic novella, *The Earthquake in Chile*, which describes the destruction of Santiago in 1647. Haruki Murakami's short fiction collection After the Quake depicts the consequences of the Kobe earthquake of 1995.

The most popular single earthquake in fiction is the hypothetical "Big One" expected of California's San Andreas Fault someday, as depicted in the novels *Richter 10* (1996), *Goodbye California* (1977), *2012* (2009) and *San Andreas* (2015) among other works. Jacob M. Appel's widely anthologized short story, *A Comparative Seismology*, features a con artist who convinces an elderly woman that an apocalyptic earthquake is imminent.

Contemporary depictions of earthquakes in film are variable in the manner in which they reflect human psychological reactions to the actual trauma that can be caused to directly afflicted families and their loved ones. Disaster mental health response research emphasizes the need to be aware of the different roles of loss of family and key community members, loss of home and familiar surroundings, loss of essential supplies and services to maintain survival. Particularly for children, the clear availability of caregiving adults who are able to protect, nourish, and clothe them in the aftermath of the earthquake, and to help them make sense of what has befallen them has been shown even more important to their emotional and physical health than the simple giving of provisions. As was observed after other disasters involving destruction and loss of life and their media depictions, recently observed in the 2010 Haiti earthquake, it is also important not to pathologize the reactions to loss and displacement or disruption of governmental administration and services, but rather to validate these reactions, to support constructive problem-solving and reflection as to how one might improve the conditions of those affected.

Aftershock

An aftershock is a smaller earthquake that occurs after a previous large earthquake, in the same area of the main shock. If an aftershock is larger than the main shock, the aftershock is redesignated as the main shock and the original main shock is redesignated as a foreshock. Aftershocks are formed as the crust around the displaced fault plane adjusts to the effects of the main shock.

Distribution of Aftershocks

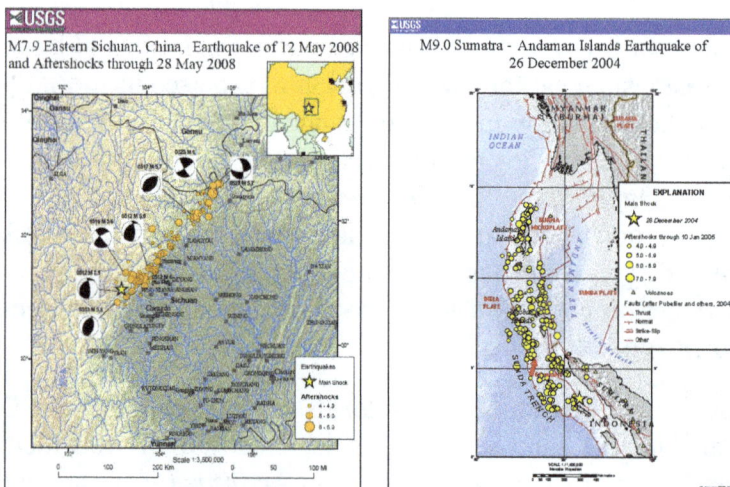

Most aftershocks are located over the full area of fault rupture and either occur along the fault plane itself or along other faults within the volume affected by the strain associated with the main shock. Typically, aftershocks are found up to a distance equal to the rupture length away from the fault plane.

The pattern of aftershocks helps confirm the size of area that slipped during the main shock. In the case of the 2004 Indian Ocean earthquake and the 2008 Sichuan earthquake the aftershock distribution shows in both cases that the epicenter (where the rupture initiated) lies to one end of the final area of slip, implying strongly asymmetric rupture propagation.

Aftershock Size and Frequency with Time

Aftershocks rates and magnitudes follow several well-established empirical laws.

Omori's Law

The frequency of aftershocks decreases roughly with the reciprocal of time after the main shock. This empirical relation was first described by Fusakichi Omori in 1894 and is known as Omori's law. It is expressed as

$$n(t) = \frac{k}{(c+t)}$$

where k and c are constants, which vary between earthquake sequences. A modified version of Omori's law, now commonly used, was proposed by Utsu in 1961.

$$n(t) = \frac{k}{(c+t)^p}$$

where p is a third constant which modifies the decay rate and typically falls in the range 0.7–1.5.

According to these equations, the rate of aftershocks decreases quickly with time. The rate of aftershocks is proportional to the inverse of time since the mainshock and this relationship can be used to estimate the probability of future aftershock occurrence. Thus whatever the probability of an aftershock are on the first day, the second day will have 1/2 the probability of the first day and the tenth day will have approximately 1/10 the probability of the first day (when p is equal to 1). These patterns describe only the statistical behavior of aftershocks; the actual times, numbers and locations of the aftershocks are stochastic, while tending to follow these patterns. As this is an empirical law, values of the parameters are obtained by fitting to data after a mainshock has occurred, and they imply no specific physical mechanism in any given case.

Båth's Law

The other main law describing aftershocks is known as Båth's Law and this states that the difference in magnitude between a main shock and its largest aftershock is approximately constant, independent of the main shock magnitude, typically 1.1–1.2 on the Moment magnitude scale.

Gutenberg–Richter Law

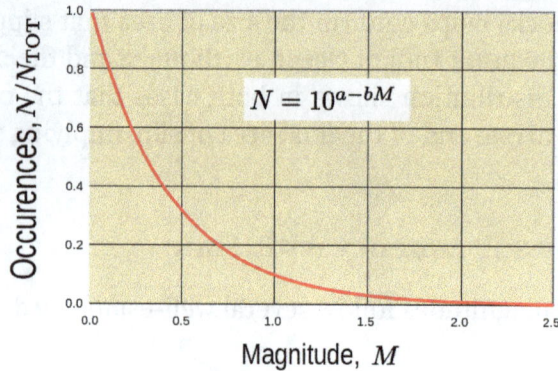

Gutenberg–Richter law for $b = 1$

Aftershock sequences also typically follow the Gutenberg–Richter law of size scaling, which refers to the relationship between the magnitude and total number of earthquakes in a region in a given time period.

$$N = 10^{a-bM}$$

Where:

- N is the number of events greater or equal to M

- M is magnitude

- a and b are constants

In summary, there are more small aftershocks and fewer large aftershocks.

Effect of Aftershocks

Aftershocks are dangerous because they are usually unpredictable, can be of a large magnitude, and can collapse buildings that are damaged from the main shock. Bigger earthquakes have more and larger aftershocks and the sequences can last for years or even longer especially when a large event occurs in a seismically quiet area; see, for example, the New Madrid Seismic Zone, where events still follow Omori's law from the main shocks of 1811–1812. An aftershock sequence is deemed to have ended when the rate of seismicity drops back to a background level; i.e., no further decay in the number of events with time can be detected.

Land movement around the New Madrid is reported to be no more than 0.2 mm (0.0079 in) a year, in contrast to the San Andreas Fault which averages up to 37 mm (1.5 in) a year across California. Aftershocks on the San Andreas are now believed to top out at 10 years while earthquakes in New Madrid are considered aftershocks nearly 200 years after the 1812 New Madrid earthquake.

Foreshocks

Some scientists have tried to use foreshocks to help predict upcoming earthquakes, having one of their few successes with the 1975 Haicheng earthquake in China. On the East Pacific Rise however, transform faults show quite predictable foreshock behaviour before the main seismic event. Reviews of data of past events and their foreshocks showed that they have a low number of aftershocks and high foreshock rates compared to continental strike-slip faults.

Modeling

Seismologists use tools such as the Epidemic-Type Aftershock Sequence model (ETAS) to study cascading aftershocks.

Foreshock

A foreshock is an earthquake that occurs before a larger seismic event (the *mainshock*) and is related to it in both time and space. The designation of an earthquake as *foreshock*, *mainshock* or aftershock is only possible after the full sequence of events has happened.

Occurrence

Foreshock activity has been detected for about 40% of all moderate to large earthquakes, and about 70% for events of M>7.0. They occur from a matter of minutes to days or even longer before the main shock, for example the 2002 Sumatra earthquake is regarded as a foreshock of the 2004 Indian Ocean earthquake with a delay of more than two years between the two events.

Some great earthquakes (M>8.0) show no foreshock activity at all, such as the M8.6 1950 India - China earthquake.

The increase in foreshock activity is difficult to quantify for individual earthquakes but becomes apparent when combining the results of many different events. From such combined observations, the increase before the mainshock is observed to be of inverse power law type. This may either indicate that foreshocks cause stress changes resulting in the mainshock or that the increase is related to a general increase in stress in the region.

Mechanics

The observation of foreshocks associated with many earthquakes suggests that they are part of a preparation process prior to nucleation. In one model of earthquake rupture, the process forms as a cascade, starting with a very small event that triggers a larger one, continuing until the main shock rupture is triggered. However, analysis of some foreshocks has shown that they tend to relieve stress around the fault. In this view, foreshocks and aftershocks are part of the same process. This is supported by an observed relationship between the rate of foreshocks and the rate of aftershocks for an event.

Earthquake Prediction

An increase in seismic activity in an area has been used as a method of predicting earthquakes, most notably in the case of the 1975 Haicheng earthquake in China, where an evacuation was triggered by an increase in activity. However, most earthquakes lack obvious foreshock patterns and this method has not proven useful, as most small earthquakes are not foreshocks, leading to probable false alarms. Earthquakes along oceanic transform faults do show repeatable foreshock behaviour, allowing the prediction of both the location and timing of such earthquakes.

Examples of Earthquakes with Foreshock Events

- The latest example of these types of earthquake was the 2016 Kumamoto earthquakes.

- The strongest earthquake of this type is the 1960 Valdivia earthquake which had a magnitude of 9.5 M_w.

Date (Fore-shock)	Mag-nitude (Fore-shock)	Flag and Country	Region	Date	Depth	Magni-tude	Intensity	Name	Deceased	Tsuna-mi
May 21, 1960 (1 day)	7.9 M_w	Chile	Araucanśa Region	May 22, 1960	35 km	9.5 M_w	XII Mercalli	1960 Valdivia earthquake	1,655	□
November 2, 2002 (2 years)	7.3 M_w	Indonesia	Sumatra	December 26, 2004	30 km	9.1 M_w	□	2004 Indian Ocean earthquake and tsunami	230,000	□
October 20, 2006 (299 days)	6.4 M_w	Peru	Ica Region	August 15, 2007	35 km	8.0 M_w	VIII Mercalli	2007 Peru earthquake	596	□
January 23, 2007 (3 months)	5.2 M_L	Chile	Aysřn Region	April 21, 2007	6 km	6.2 M_w	VII Mercalli	2007 Aysřn Fjord earthquake	10	□
March 9, 2011 (2 days)	7.3 M_w	Japan	Miyagi Prefecture	March 11, 2011	30 km	9.0 M_w	IX Mer-calli and 7 Shindo	2011 Tōhoku earthquake and tsunami	15,891	□

| March 16, 2014 (15 days) | 6.7 M_w | Chile | Tarapacō Region | April 1, 2014 | 20.1 km | 8.2 M_w | VIII Mercalli | 2014 Iquique earthquake | 7 | ▯ |
| April 14, 2016 (2 days) | 6.2 M_w | Japan | Kumamoto Prefecture | April 16, 2016 | 11 km | 7.0 M_w | IX Mercalli | 2016 Kumamoto earthquakes | 41 | X |

Note: dates are in local time

Induced Seismicity

Induced seismicity refers to typically minor earthquakes and tremors that are caused by human activity that alters the stresses and strains on the Earth's crust. Most induced seismicity is of a low magnitude. A few sites regularly have larger quakes, such as The Geysers geothermal plant in California which averaged two M4 events and 15 M3 events every year from 2004 to 2009. Results of ongoing multi-year research on induced earthquakes by the United States Geological Survey (USGS) published in 2015 suggested that most of the significant earthquakes in Oklahoma, such as the 1952 magnitude 5.7 El Reno earthquake may have been induced by deep injection of waste water by the oil industry. "Earthquake rates have recently increased markedly in multiple areas of the Central and Eastern United States (CEUS), especially since 2010, and scientific studies have linked the majority of this increased activity to wastewater injection in deep disposal wells."

Causes

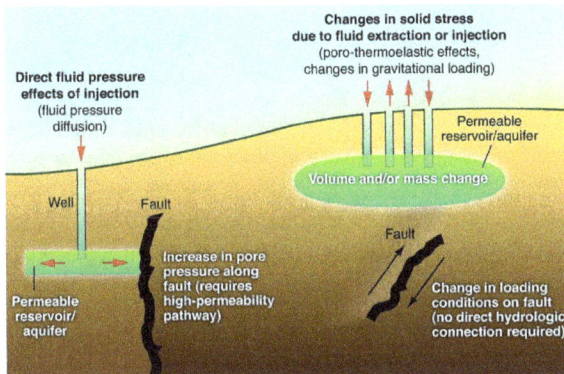

Diagram showing the effects that fluid injection and withdrawal can have on nearby faults can cause induced seismicity.

There are many ways in which induced seismicity has been seen to occur. In the past several years, some energy technologies that inject or extract fluid from the Earth, such as oil and gas extraction and geothermal energy development, have been found or suspected to cause seismic events. Some energy technologies also produce wastes that may be managed through disposal or storage by injection deep into the ground. For example, waste water from oil and gas production and carbon dioxide from a variety of industrial processes may be managed through underground injection.

Artificial Lakes

The column of water in a large and deep artificial lake alters in-situ stress along an existing fault or fracture. In these reservoirs, the weight of the water column can significantly change the stress on an underlying fault or fracture by increasing the total stress through direct loading, or decreasing the effective stress through the increased pore water pressure. This significant change in stress can lead to sudden movement along the fault or fracture, resulting in an earthquake. Reservoir-induced seismic events can be relatively large compared to other forms of induced seismicity. Though understanding of reservoir-induced seismic activity is very limited, it has been noted that seismicity appears to occur on dams with heights larger than 330 feet (100 m). The extra water pressure created by large reservoirs is the most accepted explanation for the seismic activity. When the reservoirs are filled or drained, induced seismicity can occur immediately or with a small time lag.

The first case of reservoir-induced seismicity occurred in 1932 in Algeria's Oued Fodda Dam.

The largest earthquake attributed to reservoir-induced seismicity occurred at Koyna Dam

The 6.3 magnitude 1967 Koynanagar earthquake occurred in Maharashtra, India with its epicenter, fore- and aftershocks all located near or under the Koyna Dam reservoir. 180 people died and 1,500 were left injured. The effects of the earthquake were felt 140 mi (230 km) away in Bombay with tremors and power outages.

During the beginnings of the Vajont Dam in Italy, there were seismic shocks recorded during its initial fill. After a landslide almost filled the reservoir in 1963, causing a massive flooding and around 2,000 deaths, it was drained and consequently seismic activity was almost non-existent.

On August 1, 1975, a magnitude 6.1 earthquake at Oroville, California, was attributed to seismicity from a large earth-fill dam and reservoir recently constructed and filled.

The filling of the Katse Dam in Lesotho, and the Nurek Dam in Tajikistan is an example. In Zambia, Kariba Lake may have provoked similar effects.

The 2008 Sichuan earthquake, which caused approximately 68,000 deaths, is another possible example. An article in *Science* suggested that the construction and filling of the Zipingpu Dam may have triggered the earthquake.

Some experts worry that the Three Gorges Dam in China may cause an increase in the frequency and intensity of earthquakes.

Mining

Mining leaves voids that generally alter the balance of forces in the rock, many times causing rock bursts. These voids may collapse producing seismic waves and in some cases reactivate existing faults causing minor earthquakes. Natural cavern collapse forming sinkholes would produce an essentially identical local *seismic event*.

Waste Disposal Wells

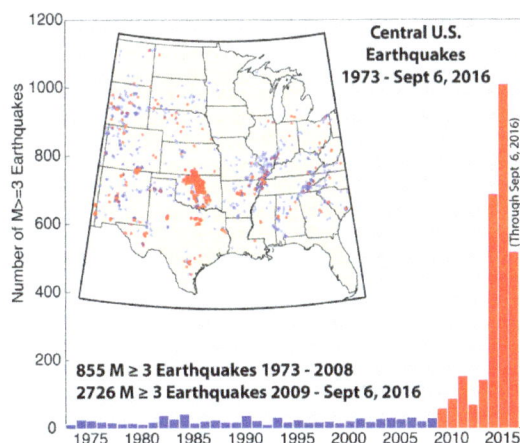

Cumulative number of earthquakes in the central U.S. The red cluster at the center of the map shows an area in and around Oklahoma which experienced the largest increase in activity since 2009.

Injecting liquids into waste disposal wells, most commonly in disposing of produced water from oil and natural gas wells, has been known to cause earthquakes. This high-saline water is usually pumped into salt water disposal (SWD) wells. The resulting increase in subsurface pore pressure can trigger movement along faults, resulting in earthquakes.

The 2011 Oklahoma earthquake near Prague, of magnitude 5.8, occurred after 20 years of injecting waste water into porous deep formations at increasing pressures and saturation. On September 3, 2016, an even stronger earthquake with a magnitude of 5.8 occurred near Pawnee, Oklahoma, followed by nine aftershocks between magnitudes 2.6 and 3.6 within 3 1/2 hours. Tremors were felt as far away as Memphis, Tennessee, and Gilbert, Arizona. Mary Fallin, the Oklahoma governor, declared a local emergency and shutdown orders for local disposal wells were ordered by the Oklahoma Corporation Commission. Results of ongoing multi-year research on induced earthquakes by the United States Geological Survey (USGS) published in 2015 suggested that most of the significant earthquakes in Oklahoma, such as the 1952 magnitude 5.5 El Reno earthquake may have been induced by deep injection of waste water by the oil industry. Prior to April 2015 however, the Oklahoma Geological Survey's position was that the quake was most likely due to natural causes and was not triggered by waste injection. This was one of many earthquakes which have affected the Oklahoma region.

Since 2009 earthquakes have become hundreds of time more common in Oklahoma with magnitude 3 events increasing from 1 or 2 per year to 1 or 2 per day. On April 21, 2015, the Oklahoma Geological Survey released a statement reversing its stance on induced earthquakes in Oklahoma: "The OGS considers it very likely that the majority of recent earthquakes, particularly those in central and north-central Oklahoma, are triggered by the injection of produced water in disposal wells."

Extraction of Fossil Fuels

Large-scale fossil fuel extraction can generate earthquakes.

Groundwater Extraction

The changes in crustal stress patterns caused by the large scale extraction of groundwater has been shown to trigger earthquakes, as in the case of the 2011 Lorca earthquake.

Geothermal Energy

Enhanced geothermal systems (EGS), a new type of geothermal power technologies that do not require natural convective hydrothermal resources, are known to be associated with induced seismicity. EGS involves pumping fluids at pressure to enhance or create permeability through the use of hydraulic fracturing techniques. Hot dry rock (HDR) EGS actively creates geothermal resources through hydraulic stimulation. Depending on the rock properties, and on injection pressures and fluid volume, the reservoir rock may respond with tensile failure, as is common in the oil and gas industry, or with shear failure of the rock's existing joint set, as is thought to be the main mechanism of reservoir growth in EGS efforts.

HDR and EGS systems are currently being developed and tested in Soultz-sous-Forêts (France), Desert Peak and the Geysers (U.S.), Landau (Germany), and Paralana and Cooper Basin (Australia). Induced seismicity events at the Geysers geothermal field in California has been strongly correlated with injection data. The test site at Basel, Switzerland, has been shut down due to induced seismic events.

Largest Events at EGS Sites Worldwide	
Site	**Maximum Magnitude**
Cerro Prieto, Baja California, Mexico	6.6
The Geysers, United States	4.6
Cooper Basin, Australia	3.7
Basel, Switzerland	3.4
Rosemanowes Quarry, United Kingdom	3.1
Soultz-sous-Forêts, France	2.9

Researchers at MIT believe that seismicity associated with hydraulic stimulation can be mitigated and controlled through predictive siting and other techniques. With appropriate management, the number and magnitude of induced seismic events can be decreased, significantly reducing the probability of a damaging seismic event.

Induced seismicity in Basel led to suspension of its HDR project. A seismic hazard evaluation was then conducted, which resulted in the cancellation of the project in December 2009.

Hydraulic Fracturing

Hydraulic fracturing is a technique in which high-pressure fluid is injected into the low-permeable reservoir rocks in order to induce fractures to increase hydrocarbon production. This process usually generates seismic events that are too small to be felt at the surface (with magnitudes ranging from -3 to 0), although several cases of larger magnitude events (M > 4) have been recorded in Canada in the unconventional resources of Alberta and British Columbia.

U.S. National Research Council Report

A 2012 report from the U.S. National Research Council examined the potential for energy technologies—including shale gas recovery, carbon capture and storage, geothermal energy production, and conventional oil and gas development—to cause earthquakes. The report found that only a very small fraction of injection and extraction activities among the hundreds of thousands of energy development sites in the United States have induced seismicity at levels noticeable to the public. However, although scientists understand the general mechanisms that induce seismic events, they are unable to accurately predict the magnitude or occurrence of these earthquakes due to insufficient information about the natural rock systems and a lack of validated predictive models at specific energy development sites.

The report noted that hydraulic fracturing has a low risk for inducing earthquakes that can be felt by people, but underground injection of wastewater produced by hydraulic fracturing and other energy technologies has a higher risk of causing such earthquakes. In addition, carbon capture and storage—a technology for storing excess carbon dioxide underground—may have the potential for inducing seismic events, because significant volumes of fluids are injected underground over long periods of time.

Cryoseism

A cryoseism, also known as an ice quake or a frost quake, is a seismic event that may be caused by a sudden cracking action in frozen soil or rock saturated with water or ice. As water drains into the ground, it may eventually freeze and expand under colder temperatures, putting stress on its surroundings. This stress builds up until relieved explosively in the form of a cryoseism.

Another type of cryoseism is a non-tectonic seismic event caused by sudden glacial movements. This movement has been attributed to a veneer of water which may pool underneath a glacier sourced from surface ice melt. Hydraulic pressure of the liquid can act as a lubricant, allowing the glacier to suddenly shift position. This type of cryoseism can be very brief, or may last for several minutes.

The requirements for a cryoseism to occur are numerous; therefore, accurate predictions are not entirely possible and may constitute a factor in structural design and engineering when constructing in an area historically known for such events. Speculation has been made between global warming and the frequency of cryoseisms.

Effects

Cryoseisms are often mistaken for minor intraplate earthquakes. Initial indications may appear similar to those of an earthquake with tremors, vibrations, ground cracking and related noises, such as thundering or booming sounds. Cryoseisms can, however, be distinguished from earthquakes through meteorological and geological conditions. Cryoseisms can have an intensity of up to VI on the Modified Mercalli Scale. Furthermore, cryoseisms often exhibit high intensity in a very localized area, in the immediate proximity of the epicenter, as compared to the widespread effects of an earthquake. Due to lower-frequency vibrations of cryoseisms, some seismic monitoring stations may not record their occurrence. Although cryoseisms release less energy than most tectonic events, they can still cause damage or significant changes to an affected area.

Some reports have indicated the presence of "distant flashing lights" before or during a cryoseism, possibly because of electrical changes when rocks are compressed. Cracks and fissures may also appear as surface areas contract and split apart from the cold. The sometime superficial to moderate occurrences may range from a few centimeters to several kilometers long, with either singular or multiple linear fracturing and vertical or lateral displacement possible.

Occurrences

Location

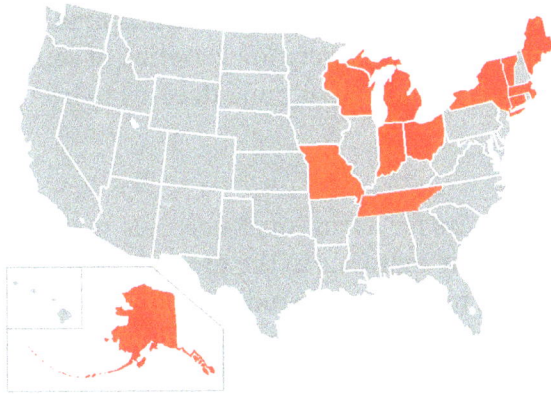

US States with reported cryoseisms.

Geocryological processes were identified as a possible cause of tremors as early as 1818. In the United States, such events have been reported throughout the Midwestern, Northern and Northeastern United States.

Cryoseisms also occur in Canada, especially along the Great Lakes/St. Lawrence corridor, where winter temperatures can shift very rapidly. They have surfaced in Ontario, Quebec and the Maritime Provinces. They are also observed in Calgary.

Glacier-related cryoseism phenomena have been reported in Alaska, Greenland, Iceland (Grímsvötn)., Ross Island, and the Antarctic Prince Charles Mountains.

Reports

Southern Ontario experienced numerous, albeit unverified, frost quakes on December 24, 2013, and more severe, waking a great many from sleep, on December 30, 2013 at 3:00 AM and later at 7:00 AM, as a result of the 2013 Central and Eastern Canada ice storm.

The towns of Cornwall, Middlebury, Ripton, and Weybridge, Vermont experienced the same phenomenon after the same weather conditions with the loudest occurring around 4:50am on December 25. Essex County in New York experienced cryoseisms as late as December 31, 2013. Numerous frost quakes again occurred in the Greater Toronto Area around 2:00 AM January 3, 2014, and January 7, 2014 overnight. Many cryoseisms were also reported near Lac-Mégantic, in the Eastern Townships (Québec, Canada), in the last days of December 2013.

On March 4, 2014 the city of Calgary experienced a loud sound at 5:00 PM and many people are attributing this to cryoseism.

The West Island of Montreal experienced an ice quake at approximately 2:30 AM EST on January 6, 2015.

Wichita, Kansas reportedly experienced a loud boom on Saturday, January 17, 2016. It was the first recorded ice quake in Kansas.

During a January 2016 record cold spell several unusual loud bangs and tremors were reported around Finland, which has a very low baseline of seismological activity. One of the events was registered seismologically and deemed to be a frost quake.

Precursors

There are four main precursors for a frost quake cryoseism event to occur: (1) a region must be susceptible to cold air masses, (2) the ground must undergo saturation from thaw or liquid precipitation prior to an intruding cold air mass, (3) most frost quakes are associated with minor snow cover on the ground without a significant amount of snow to insulate the ground (i.e., less than 6 inches), and (4) a rapid temperature drop from approximately freezing to near or below zero degrees Fahrenheit, which ordinarily occurred on a timescale of 16 to 48 hours.

Cryoseisms typically occur when temperatures rapidly decrease from above freezing to subzero, and are more than likely to occur between midnight and dawn (during the coldest parts of night). [However, due to the permanent nature of glacial ice, glacier-related cryoseisms may also occur in the warmer months of summer.] In general, cryoseisms may occur 3 to 4 hours after significant changes in temperature. Perennial or seasonal frost conditions involved with cryoseisms limit these events to temperate climates that experience seasonal variation with subzero winters. Additionally, the ground must be saturated with water, which can be caused by snowmelt, rain, sleet or flooding. Geologically, areas of permeable materials like sand or gravel, which are susceptible to frost action, are likelier candidates for cryoseisms. Following large cryoseisms, little to no seismic activity will be detected for several hours, indicating that accumulated stress has been relieved.

Submarine Earthquake

A submarine, undersea, or underwater earthquake is an earthquake that occurs underwater at the bottom of a body of water, especially an ocean. They are the leading cause of tsunamis. The magnitude can be measured scientifically by the use of the Richter scale and the intensity can be assigned using the Mercalli intensity scale.

Understanding plate tectonics helps to explain the cause of submarine earthquakes. The Earth's surface or lithosphere comprises tectonic plates which average approximately 50 miles in thickness, and are continuously moving very slowly upon a bed of magma in the asthenosphere and inner mantle. The plates converge upon one another, and one subducts below the other, or, where there is only shear stress, move horizontally past each other. Little movements called fault creep are minor and not measurable.

The plates meet with each other, and if rough spots cause the movement to stop at the edges, the motion of the plates continue. When the rough spots can no longer hold, the sudden release of the built-up motion releases, and the sudden movement under the sea floor causes a submarine earthquake. This area of slippage both horizontally and vertically is called the epicenter, and has the highest magnitude, and causes the greatest damage.

As with a continental earthquake the severity of the damage is not often caused by the earthquake at the rift zone, but rather by events which are triggered by the earthquake. Where a continental earthquake will cause damage and loss of life on land from fires, damaged structures, and flying objects; a submarine earthquake alters the sea bed floor, resulting in a series of waves, and depending on the length and magnitude of the earthquake, huge tidal waves and tsunamis, which bear down on coastal cities causing property damages and loss of life.

Submarine earthquakes can also damage submarine communications cables, leading to widespread disruption of the Internet and international telephone network in those areas. This is particularly common in Asia, where many submarine links cross submarine earthquake zones such as the Pacific Ring of Fire.

Tectonic Plate Boundaries

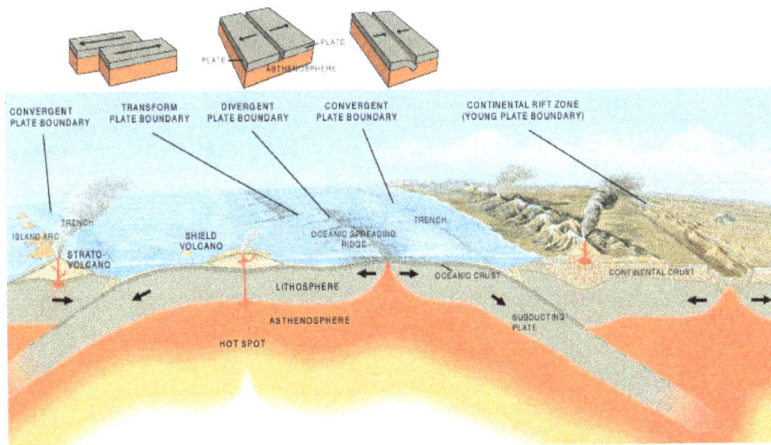

Different kinds of boundaries

The different ways in which tectonic plates rub against each other under the ocean or sea floor to create submarine earthquakes. The type of friction created may be due to the characteristic of the geologic fault or the plate boundary as follows. Some of the main areas of large tsunami producing submarine earthquakes are the Pacific Ring of Fire and the Great Sumatran fault.

Convergent Plate Boundary

The older, and denser plate moves below the lighter plate. The further down it moves, the hotter it becomes, until finally melting altogether at the asthenosphere and inner mantle and the crust is actually destroyed. The location where the two oceanic plates actually meet become deeper and deeper creating trenches with each successive action. There is an interplay of various densities of lithosphere rock, asthenosphere magma, cooling ocean water and plate movement for example the Pacific Ring of Fire. Therefore, the site of the sub oceanic trench will be a site of submarine earthquakes; for example the Mariana Trench, Puerto Rico Trench, and the volcanic arc along the Great Sumatran fault.

Transform Plate Boundary

A transform-fault boundary, or simply a transform boundary is where two plates will slide past each other, and the irregular pattern of their edges may catch on each other. The lithosphere is neither added to from the asthenosphere nor is it destroyed as in convergent plate action. For example, along the San Andreas fault strike-slip fault zone, the Pacific Tectonic Plate has been moving along at about 5 cm/yr in a northwesterly direction, whereas the North American Plate is moving south-easterly.

Divergent Plate Boundary

Rising convection currents occur where two plates are moving away from each other. In the gap, thus produced hot magma rises up, meets the cooler sea water, cools, and solidifies, attaching to either or both tectonic plate edges creating an oceanic spreading ridge. When the fissure again appears, again magma will rise up, and form new lithosphere crust. If the weakness between the two plates allows the heat and pressure of the asthenosphere to build over a large amount of time, a large quantity of magma will be released pusing up on the plate edges and the magma will solidify under the newly raised plate edges, see formation of a submarine volcano. If the fissure is able to come apart because of the two plates moving apart, in a sudden movement, an earthquake tremor may be felt for example at the Mid-Atlantic Ridge between North America and Africa.

List of Major Submarine Earthquakes

The following is a list of major submarine earthquakes.

Date	Event	Location	Estimated moment magnitude (M_w)	Notes
March 11, 2011	2011 Tōhoku earthquake	The epicenter is 130 kilometers (81 mi) off the east coast of the Oshika Peninsula, Tōhoku, with the hypocenter at a depth of 32 km (20 mi).	9.0	This is strongest known earthquake to hit Japan

December 26, 2006	2006 Hengchun earthquakes	The epicenter is off the south-west coast of Taiwan, in the Luzon Strait, which connects the South China Sea with the Philippine Sea.	7.1	
December 26, 2004	2004 Indian Ocean earthquake	The epicenter is off the north-western coast of Sumatra, Indonesia.	9.3	This is the third largest earthquake in recorded history and generated massive tsunamis, which caused widespread devastation when they hit land, leaving an estimated 230,000 people dead in countries around the Bay of Bengal and the Indian Ocean.
May 4, 1998		A part of the island of Yo-naguni was destroyed by a submarine earthquake.		
May 22, 1960	1960 Valdivia earthquake	The epicenter is off the coast of South Central Chile.	9.5	This is the strongest earthquake ever recorded.
December 20, 1946	1946 Nankaido earthquake	The epicenter is off the southern coast of Kii Peninsula and Shikoku, Japan.	8.1	
December 7, 1944	1944 Tōnankai earthquake	The epicenter is about 20 km off the coast of the Shima Peninsula in Japan.	8.0	
November 18, 1929	1929 Grand Banks earthquake	The epicenter is at Grand Banks, off the south coast of Newfoundland in the Atlantic Ocean.	7.2	
June 15, 1896	1896 Sanriku earthquake	The epicenter is off the Sanriku coast of northeastern Honshū, Japan.	8.5	
April 4, 1771		The epicenter is near Yaeyama Islands in Okinawa, Japan.	7.4	

Jan-uary 26, 1700	1700 Cascadia earth-quake	The epicenter is offshore from Vancouver Island to northern California.	~9.0		This is one of the largest earthquakes on record.

Seismic Site Effects

Seismic site effects are related to the amplification of seismic waves in surficial geological layers. The surface ground motion may be strongly amplified if the geological conditions are unfavourable (e.g. sediments). The damages due to an earthquake may thus be aggravated as in the case of the 1985 Mexico City earthquake. For alluvial basins, we may shake a bowl of jelly to model the phenomenon at a small scale.

This article defines site effects first, presents the 1985 Mexico City earthquake, describes the theoretical analysis of the phenomenon (through mechanical waves) and details several research results on seismic site effects in Caracas.

Definition of the Phenomenon

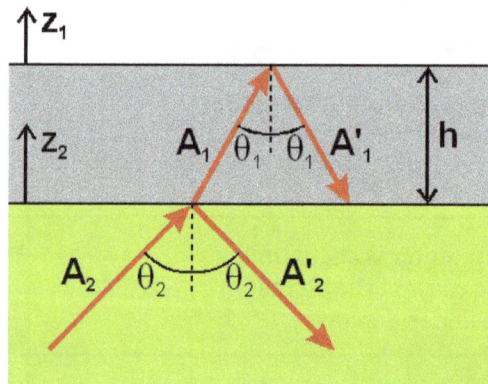

Fig.1 : Seismic site effects / wave amplification in a horizontal layer (SH-waves): various wavefields.

When propagating, the seismic waves are reflected and refracted at the interface between the various geological layers (Fig.1).

The example of Figure 1 depicts the seismic wave amplification in horizontal geological layers. We consider a homogeneous elastic half-space (in green) over which an elastic alluvial layer of constant thickness h is located (in gray). A shear wave (SH) of amplitude A_2 reaches the interface between the half-space and the alluvial layer with an incidence θ_2. It thus generates:

- a reflected wave in the half-space with amplitude A_2' and incidence θ_2

- a refracted wave in the surficial layer with amplitude A_1 and incidence θ_1

The refracted wave originates a reflected wave when reaching the free surface libre; its amplitude and incidence are denoted A_1 and θ_1 respectively. This latter wave will be reflected and refracted several times at the base and the top of the surficial layer. If the layer is softer than the half-space, the surface motion amplitude can be larger than A_2 thus leading to the *amplification of seismic waves* or *seismic site effetcs*. When the geological interfaces are not horizontal, it also possible to study seismic site effetcs by considering the *basin effects* due to the complexe geometry of the alluvial filling

In this article, we propose several examples of seismic site effects (observed or simulated during large earthquakes) as well as a theoretical analysis of the amplification phenomenon.

Example: Site Effects in Mexico City (1985)

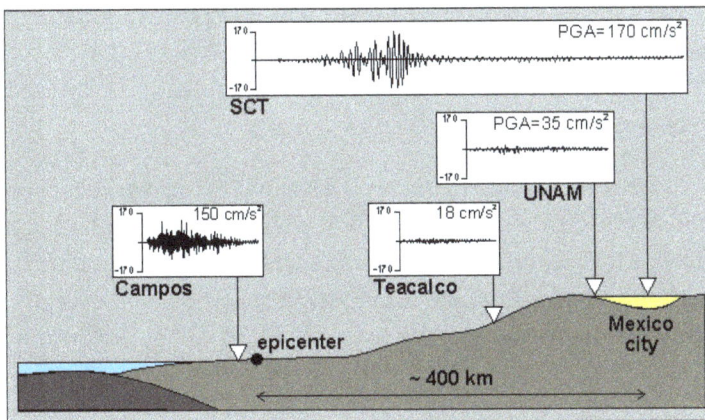

Fig.2 : Site effects in Mexico city: recordings from the 1985 earthquake

Seismic site effects have been first evidenced during the 1985 Mexico City earthquake. The earthquake epicenter was located along the Pacific Coast (several hundreds kilometers from Mexico-City), the seismic shaking was however extremely strong leading to very large damages.

Figure 2 displays the recordings performed at different distances from the epicenter during the earthquake sequence. The acceleration amplitude measured at different distances changes drastically:

- *Campos* station: this station is located very close to the epicenter and recorded a maximum acceleration of $150 \; cm/s^2$,

- *Teacalco* station: this station is located at more than 200 km from the epicenter and recorded a much lower acceleration (about $18 cm/s^2$). This amplitude decay is due to the wave attenuation during the propagation process: geometrical attenuation due to the expansion of the wavefront and material (or instrinsic) attenuation due to the energy dissipation within the medium (e.g. grains friction),

- *UNAM* station: this station is located at more than 300 km from the epicenter and recorded a maximum acceleration of $35\ cm/s^2$, larger than that recorded at the *Teacalco* station,

- *SCT* station: this station is located in Mexico City at approximately 400 km from the epicenter and recorded a very strong maximum acceleration (about $170\ cm/s^2$).

We may notice that the acceleration amplitude strongly decreases first and then increases when the seismic waves reach the alluvial deposit on which Mexico City has been founded.

Theoretical Analysis of Seismic Site Effects: Horizontal Layering

In case of horizontal soil layering (constant thickness, cf Fig.1), we may analyze seismic site effects theoretically. One considers a shear wave (*SH*) (i.e. polarized perpendicularly to the figure) reflected and refracted wave at the interface between both media and reflected at the free surface.

Considering Fig.1, we may analyze the propagation of the various waves in the sedimentary layer ($i=1$) and in the half-space ($i=2$). Assuming both media as linear elastic and writing the continuity conditions at the interface (displacement and traction) as well as the free surface conditions, we may determine the spectral ratio $\bar{T}(\omega)$ between the surface motion and the motion at the top of the half-space without any sedimentary layer:

$$\bar{T}(\omega) = \frac{2A_1}{2A_2} = \frac{1}{\cos k_{z_1} h + i\bar{\chi} \sin k_{z_1} h}$$

where $k_{z_1} = \dfrac{\omega \theta_i}{V_{S_i}}$; $\bar{\chi} = \sqrt{\dfrac{\mu_1 \rho_1}{\mu_2 \rho_2}} \dfrac{\cos \theta_1}{\cos \theta_2}$ and :

- h is the layer thickness,

- θ_i is the wave incidence in layer i,

- ρ_i is the mass density in layer i,

- μ_i is the shear modulus in layer i,

- k_{z_1} is the vertical wave number in layer 1,

- $V_{S_i} = \sqrt{\dfrac{\mu_i}{\rho_i}} is$ is the shear wave velocity.

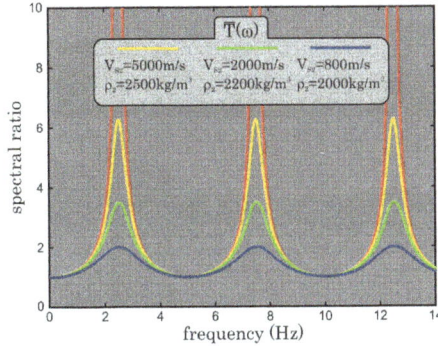

Fig. 3: Seismic site effects in a single sedimentary layer (SH waves): spectral ratio for various layer/bedrock velocity ratios

Fig.3 displays the variations of the spectral ratio $\bar{T}$ with respect to frequency for different mechanical features of the half-space (with $V_{S_1} = 200\ m/s$ for the sedimentary layer). We notice that the motion amplification may be very strong at certain frequencies. The amplification level depends on the velocity contrast $\bar{\chi}$ and takes the following maximum values:

- $|\bar{T}_{max}| = 2$ for $V_{S_2} = 800\ m/s$ (blue curve),

- $|\bar{T}_{max}| = 3.5$ for $V_{S_2} = 2000\ m/s$ (green curve),

- $|\bar{T}_{max}| = 6$ for $V_{S_2} = 5000\ m/s$ (yellow curve).

The red curve corresponds to a large velocity contrast between the layer and the half-space ($\bar{\chi} = 1$); the amplification is thus very large. As displayed in Fig.3, the maximum amplification is reached at certain frequencies corresponding to the resonance of the sedimentary layer. The fundamental frequency of the layer (or 1st resonance frequency) may be easily calculated under the form $f_0 = \dfrac{V_{S_1}}{4h}$: . The fundamental mode thus corresponds to a quarter wavelength resonance.

When the sedimentary layers are not horizontal (e.g. sedimentary basin), the analysis is more complex since the surface waves generated by the lateral heterogeneities (e.g. basin edges) should be accounted for. In such cases, it is possible to perform empirical studies but also theoretical analyses for simple geometries or numerical simulations for more complex cases.

Seismic Site Effects in Sedimentary Basins: The Case of Caracas

In sedimentary basins, site effects also lead to the generation of surface waves at the basin edges. This phenomenon may significantly strengthen the amplification of the seismic motion. The aggravation of the amplification level when compared to the case of horizontal layering may be up to a factor of 5 or 10. It depends on the velocity contrast between the layers and the geometry of the basin. Such phenomena are named *basin effects* and we may consider the analogy with the vibrations in a bowl of jelly.

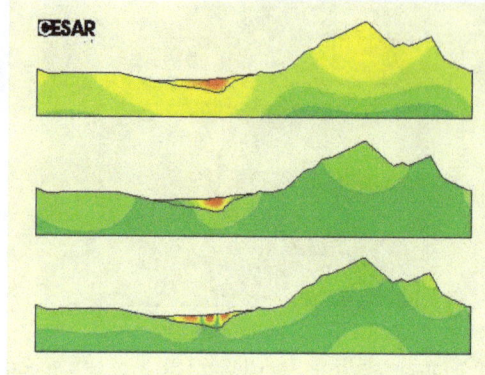

Fig.4: Seismic site effects in Caracas (BEM simulations in the frequency domain).

The theoretical analysis of site effects in canyons or semi-circular sedimentary basins has been performed through semi-analytical methods in the early 80's. Recent numerical simulations allowed the analysis of site effects in ellipsoidal sedimentary basins. Depending on the basin geometry, the aggravation of site effects is different from that of the horizontally layered case.

When the mechanical properties of the sedimentary basin are known, we may simulate site effects numerically. Figure 4 depicts the amplification phenomenon for the city of Caracas. The amplification level of a plane wave (SH) is computed by the Boundary Element Method in the frequency domain. Each color map displays the amplification level A_0 at a given frequency f_0:

- top: $f_0 = 0.3 Hz$; $A_0 = 2.53$. Site effects due to the topography clearly occur at the top of the hill (right). Nevertheless, site effects due to the sedimentary basin lead to a larger amplification.

- middle: $f_0 = 0.4 Hz$; $A_0 = 8.83$. Topographical site effects are negligible when compared to that due to the basin (4 times larger than at 0.3 Hz).

- bottom: $f_0 = 0.6 Hz$; $A_0 = 7.11$. Site effects in the basin are of the same order than at 0.4 Hz but we notice a much shorter wavelength.

Numerous geological sites have been investigated by various researchers for weak earthquakes as well as for strong ones (cf synthesis). In the latter case, it is necessary to account for the nonlinear behavior of the soil under large loadings or even the soil liquefaction which may lead to the soil failure.

Earthquake Prediction

Earthquake prediction is a branch of the science of seismology concerned with the specification of the time, location, and magnitude of future earthquakes within stated limits, and particularly "the determination of parameters for the next strong earthquake to occur in a region. Earthquake prediction is sometimes distinguished from earthquake forecasting, which can be defined as the probabilistic assessment of general earthquake hazard, including the frequency and magnitude of damaging earthquakes in a given area over years or decades. Prediction can be further distinguished from earthquake warning systems, which upon detection of an earthquake, provide a real-time warning of seconds to neighboring regions that might be affected.

In the 1970s, scientists were optimistic that a practical method for predicting earthquakes would soon be found, but by the 1990s continuing failure led many to question whether it was even possible. Demonstrably successful predictions of large earthquakes have not occurred and the few claims of success are controversial. Extensive searches have reported many possible earthquake precursors, but, so far, such precursors have not been reliably identified across significant spatial and temporal scales. While part of the scientific community hold that, taking into account non-seismic precursors and given enough resources to study them extensively, prediction might be possible, most scientists are pessimistic and some maintain that earthquake prediction is inherently impossible.

Evaluating Earthquake Predictions

Predictions are deemed significant if they can be shown to be successful beyond random chance. Therefore, methods of statistical hypothesis testing are used to determine the probability that an earthquake such as is predicted would happen anyway (the null hypothesis). The predictions are then evaluated by testing whether they correlate with actual earthquakes better than the null hypothesis.

Option:	If Quake:	If No Quake:
Alarm	**Great losses, mitigated by** preparations (cost of alarm incidental).	**False alarm:** cost of alarm, panic and economic disruption. Multiple instances?
The Bar		but increases the cost of false alarms.
Lowering the bar..	reduces odds of losses	
No Alarm	**Great losses, worsened by** being caught off–guard.	**Normal:** no losses, no disruption, no cost of alarm.

The Dilemma: To Alarm? or Not to Alarm?

In many instances, however, the statistical nature of earthquake occurrence is not simply homogeneous. Clustering occurs in both space and time. In southern California about 6% of M≥3.0 earthquakes are "followed by an earthquake of larger magnitude within 5 days and 10 km." In central Italy 9.5% of M≥3.0 earthquakes are followed by a larger event within 30 km and 48 hours. While such statistics are not satisfactory for purposes of prediction (giving ten to twenty false alarms for each successful prediction) they will skew the results of any analysis that assumes that earthquakes occur randomly in time, for example, as realized from a Poisson process. It has been shown that a "naive" method based solely on clustering can successfully predict about 5% of earthquakes; slightly better than chance.

As the purpose of short-term prediction is to enable emergency measures to reduce death and destruction, failure to give warning of a major earthquake, that does occur, or at least an adequate evaluation of the hazard, can result in legal liability, or even political purging. But warning of an earthquake that does not occur also incurs a cost: not only the cost of the emergency measures themselves, but of civil and economic disruption. False alarms, including alarms that are cancelled, also undermine the credibility, and thereby the effectiveness, of future warnings. The acceptable trade-off between missed quakes and false alarms depends on the societal valuation of these outcomes. The rate of occurrence of both must be considered when evaluating any prediction method.

In a 1997 study of the cost-benefit ratio of earthquake prediction research in Greece, Stathis Stiros suggested that even a (hypothetical) excellent prediction method would be of questionable social utility, because "organized evacuation of urban centers is unlikely to be successfully accomplished", while "panic and other undesirable side-effects can also be anticipated." He found that earthquakes kill less than ten people per year in Greece (on average), and that most of those fatalities occurred in large buildings with identifiable structural issues. Therefore, Stiros stated that it would be much more cost-effective to focus efforts on identifying and upgrading unsafe buildings. Since the death toll on Greek highways is more than 2300 per year on average, he argued that more lives would also be saved if Greece's entire budget for earthquake prediction had been used for street and highway safety instead.

Difficulty or Impossibility

All predictions of the future can be to some extent successful by chance.

—Mulargia & Gasperini 1992

Earthquake prediction may be intrinsically impossible. It has been argued that the Earth is in a state of self-organized criticality "where any small earthquake has some probability of cascading into a large event". It has also been argued on decision-theoretic grounds that prediction of major earthquakes is impossible. However, these theories and their implication that earthquake prediction is intrinsically impossible have been disputed.

Prediction Methods

Earthquake prediction is an immature science—it has not yet led to a successful prediction of an earthquake from first physical principles. Research into methods of prediction therefore focus on empirical analysis, with two general approaches: either identifying distinctive *precursors* to earthquakes, or identifying some kind of geophysical *trend* or pattern in seismicity that might precede a large earthquake. Precursor methods are pursued largely because of their potential utility for short-term earthquake prediction or forecasting, while 'trend' methods are generally thought to be useful for forecasting, long term prediction (10 to 100 years time scale) or intermediate term prediction (1 to 10 years time scale).

Precursors

An earthquake precursor is an anomalous phenomenon that might give effective warning of an impending earthquake. Reports of these – though generally recognized as such only after the event – number in the thousands, some dating back to antiquity. There have been around 400 reports of possible precursors in scientific literature, of roughly twenty different types, running the gamut from aeronomy to zoology. None have been found to be reliable for the purposes of earthquake prediction.

In the early 1990, the IASPEI solicited nominations for a Preliminary List of Significant Precursors. Forty nominations were made, of which five were selected as possible significant precursors, with two of those based on a single observation each.

After a critical review of the scientific literature the *International Commission on Earthquake Forecasting for Civil Protection* (ICEF) concluded in 2011 there was "considerable room for methodological improvements in this type of research." In particular, many cases of reported precursors are contradictory, lack a measure of amplitude, or are generally unsuitable for a rigorous statistical evaluation. Published results are biased towards positive results, and so the rate of false negatives (earthquake but no precursory signal) is unclear. Seiya Uyeda has suggested that a significant problem is that the vast majority of seismic data gathering stations are sensitive only to seismic and geodetic data, and are not configured to detect geo-electromagnetic signals or other precursors. Therefore, the required data for comprehensive and definitive studies does not exist.

Animal Behavior

For centuries there have been anecdotal accounts of anomalous animal behavior preceding and associated with earthquakes. In cases where animals display unusual behavior some tens of seconds prior to a quake, it has been suggested they are responding to the P-wave. These travel through the ground about twice as fast as the S-waves that cause most severe shaking. They predict not the earthquake itself — that has already happened — but only the imminent arrival of the more destructive S-waves.

It has also been suggested that unusual behavior hours or even days beforehand could be triggered by foreshock activity at magnitudes that most people do not notice. Another confounding factor of accounts of unusual phenomena is skewing due to "flashbulb memories": otherwise unremarkable details become more memorable and more significant when associated with an emotionally powerful event such as an earthquake. A study that attempted to control for these kinds of factors found an increase in unusual animal behavior (possibly triggered by foreshocks) in one case, but not in four other cases of seemingly similar earthquakes.

Changes in Vp/Vs

V_p is the symbol for the velocity of a seismic "P" (primary or pressure) wave passing through rock, while V_s is the symbol for the velocity of the "S" (secondary or shear) wave. Small-scale laboratory experiments have shown that the ratio of these two velocities – represented as V_p/V_s – changes when rock is near the point of fracturing. In the 1970s it was considered a likely breakthrough when Russian seismologists reported observing such changes in the region of a subsequent earthquake. This effect, as well as other possible precursors, has been attributed to dilatancy, where rock stressed to near its breaking point expands (dilates) slightly.

Study of this phenomena near Blue Mountain Lake in New York State led to a successful prediction in 1973. However, additional successes have not followed, and it has been suggested that the prediction was a fluke. A V_p/V_s anomaly was the basis of a 1976 prediction of a M 5.5 to 6.5 earthquake near Los Angeles, which failed to occur. Other studies relying on quarry blasts (more precise, and repeatable) found no such variations; and an alternative explanation has been reported for such variations as have been observed. Geller (1997) noted that reports of significant velocity changes have ceased since about 1980.

Radon Emissions

Most rock contains small amounts of gases that can be isotopically distinguished from the normal atmospheric gases. There are reports of spikes in the concentrations of such gases prior to a major earthquake; this has been attributed to release due to pre-seismic stress or fracturing of the rock. One of these gases is radon, produced by radioactive decay of the trace amounts of uranium present in most rock.

Radon is useful as a potential earthquake predictor because it is radioactive and thus easily detected, and its short half-life (3.8 days) makes radon levels sensitive to short-term fluctuations. A 2009 review found 125 reports of changes in radon emissions prior to 86 earthquakes since 1966. But as the ICEF found in its review, the earthquakes with which these changes are supposedly linked were up to a thousand kilometers away, months later, and at all magnitudes. In some cases the anomalies were observed at a distant site, but not at closer sites. The ICEF found "no significant correlation". Another review concluded that in some cases changes in radon levels preceded an earthquake, but a correlation is not yet firmly established.

Electromagnetic Variations

Various attempts have been made to identify possible pre-seismic indications in electrical, electric-resistive, or magnetic phenomena.

VAN Method

The VAN method is named after the authors, P. Varotsos, K. Alexopoulos and K. Nomicos of the National and Capodistrian University of Athens. In a 1981 paper they claimed that by measuring ultra low frequency geoelectric voltages – what they called "seismic electric signals" (SES) – they could predict earthquakes. In 1980s the VAN team published more on their research and in 1990s they claimed they were able to predict earthquakes larger than magnitude 5, within 100 km of the epicentral location, within 0.7 units of magnitude up to several weeks beforehand, and published a series of predictions.

Objections have been raised that the physics of the VAN method is not possible and the analysis of the propagation properties of SES in the Earth's crust claimed that it would have been impossible for signals with the amplitude reported by VAN to have been transmitted over the several hundred kilometers distances from the epicenter to the monitoring station. It was also claimed that VAN's publications do not account for (i.e. identify and eliminate) possible sources of electromagnetic interference (EMI).

Taken as a whole, the VAN method has been criticized as lacking consistency in the statistical testing of the validity of their hypotheses. In particular, there has been some contention regarding the catalog of actual seismic events to use in vetting predictions, and of the predicted magnitudes. A critical study found that, because of such ambiguities, a set of 22 claims of successful prediction by VAN were actually 74% false, 9% correlated at random and for 14% the correlation was uncertain. Two major reviews of VAN appeared in 1996.

Since 2001, the VAN group has introduced a concept they call "natural time", applied to the analysis of their precursors. Initially it is applied on SES to distinguish them from noise and relate them to a possible impending earthquake. In case of verification (classification as "SES activity"), natural time analysis is additionally applied to the general subsequent seismicity of the area associated with the SES activity, in order to improve the time parameter of the prediction. The method treats earthquake onset as a critical phenomenon.

Background Magnetic Field Noise Disturbances

After the 1989 Loma Prieta earthquake occurred, a group led by Antony C. Fraser-Smith of Stanford University reported that the event was preceded by disturbances in background magnetic field noise as measured by a sensor placed in Corralitos, California, about 4.5 miles (7 km) from the epicenter. From 5 October, they reported a substantial increase in noise in the frequency range 0.01–10 Hz. The measurement instrument was a single-axis search-coil magnetometer that was being used for low frequency research. Precursory increases of noise apparently started a few days before the earthquake, with noise in the range .01–.5 Hz rising to exceptionally high levels about three hours before the earthquake. Though this pattern gave scientists new ideas for research into potential precursors to earthquakes, and the Fraser-Smith et al. report remains one of the most frequently cited examples of a specific earthquake precursor, more recent studies have cast doubt on the connection, attributing the Corralitos signals to either unrelated magnetic disturbance or, even more simply, to sensor-system malfunction.

Trends

Instead of watching for anomalous phenomena that might be precursory signs of an impending earthquake, other approaches to predicting earthquakes look for trends or patterns that lead to an earthquake. As these trends may be complex and involve many variables, advanced statistical techniques are often needed to understand them, therefore these are sometimes called statistical methods. These approaches also tend to be more probabilistic, and to have larger time periods, and so merge into earthquake forecasting.

Elastic Rebound

Even the stiffest of rock is not perfectly rigid. Given a large force (such as between two immense tectonic plates moving past each other) the earth's crust will bend or deform. According to the elastic rebound theory of Reid (1910), eventually the deformation (strain) becomes great enough that something breaks, usually at an existing fault. Slippage along the break (an earthquake) allows the rock on each side to rebound to a less deformed state. In the process energy is released in various forms, including seismic waves. The cycle of tectonic force being accumulated in elastic deformation and released in a sudden rebound is then repeated. As the displacement from a single earthquake ranges from less than a meter to around 10 meters (for an M 8 quake), the demonstrated existence of large strike-slip displacements of hundreds of miles shows the existence of a long running earthquake cycle.

Characteristic Earthquakes

The most studied earthquake faults (such as the Nankai megathrust, the Wasatch fault, and the San Andreas fault) appear to have distinct segments. The *characteristic earthquake* model postulates that earthquakes are generally constrained within these segments. As the lengths and other properties of the segments are fixed, earthquakes that rupture the entire fault should have similar characteristics. These include the maximum magnitude (which is limited by the length of the rupture), and the amount of accumulated strain needed to rupture the fault segment. Since continuous plate motions cause the strain to accumulate steadily, seismic activity on a given segment should be dominated by earthquakes of similar characteristics that recur at somewhat regular intervals. For a given fault segment, identifying these characteristic earthquakes and timing their recurrence rate (or conversely return period) should therefore inform us about the next rupture; this is the approach generally used in forecasting seismic hazard. UCERF3 is a notable example of such a forecast, prepared for the state of California. Return periods are also used for forecasting other rare events, such as cyclones and floods, and assume that future frequency will be similar to observed frequency to date.

The idea of characteristic earthquakes was the basis of the Parkfield prediction: fairly similar earthquakes in 1857, 1881, 1901, 1922, 1934, and 1966 suggested a pattern of breaks every 21.9 years, with a standard deviation of ±3.1 years. Extrapolation from the 1966 event led to a prediction of an earthquake around 1988, or before 1993 at the latest (at the 95% confidence interval). The appeal of such a method is that the prediction is derived entirely from the *trend*, which supposedly accounts for the unknown and possibly unknowable earthquake physics and fault parameters. However, in the Parkfield case the predicted earthquake did not occur until 2004, a decade late. This seriously undercuts the claim that earthquakes at Parkfield are quasi-periodic, and suggests the individual events differ sufficiently in other respects to question whether they have distinct characteristics in common.

Further research into the Parkfield seismic data revealed that several 4.0 earthquakes had reduced the stresses on the northwest portion of the Parkfield segment, causing it to skip generating the predicted 6.0 earthquake.

The failure of the Parkfield prediction has raised doubt as to the validity of the characteristic earthquake model itself. Some studies have questioned the various assumptions, including the key one that earthquakes are constrained within segments, and suggested that the "characteristic earthquakes" may be an artifact of selection bias and the shortness of seismological records (relative to earthquake cycles). Other studies have considered whether other factors need to be considered, such as the age of the fault. Whether earthquake ruptures are more generally constrained within a segment (as is often seen), or break past segment boundaries (also seen), has a direct bearing on the degree of earthquake hazard: earthquakes are larger where multiple segments break, but in relieving more strain they will happen less often.

Seismic Gaps

At the contact where two tectonic plates slip past each other every section must eventually slip, as (in the long-term) none get left behind. But they do not all slip at the same time; different sections will be at different stages in the cycle of strain (deformation) accumulation and sudden rebound. In the seismic gap model the "next big quake" should be expected not in the segments where recent seismicity has relieved the strain, but in the intervening gaps where the unrelieved strain is the greatest. This model has an intuitive appeal; it is used in long-term forecasting, and was the basis of a series of circum-Pacific (Pacific Rim) forecasts in 1979 and 1989–1991.

However, some underlying assumptions about seismic gaps are now known to be incorrect. A close examination suggests that "there may be no information in seismic gaps about the time of occurrence or the magnitude of the next large event in the region"; statistical tests of the circum-Pacific forecasts shows that the seismic gap model "did not forecast large earthquakes well". Another study concluded that a long quiet period did not increase earthquake potential.

Seismicity patterns

Various heuristically derived algorithms have been developed for predicting earthquakes. Probably the most widely known is the M8 family of algorithms (including the RTP method) developed under the leadership of Vladimir Keilis-Borok. M8 issues a "Time of Increased Probability" (TIP) alarm for a large earthquake of a specified magnitude upon observing certain patterns of smaller earthquakes. TIPs generally cover large areas (up to a thousand kilometers across) for up to five years. Such large parameters have made M8 controversial, as it is hard to determine whether any hits that happened were skillfully predicted, or only the result of chance.

M8 gained considerable attention when the 2003 San Simeon and Hokkaido earthquakes occurred within a TIP. In 1999, Keilis-Borok's group published a claim to have achieved statistically significant intermediate-term results using their M8 and MSc models, as far as world-wide large earthquakes are regarded. However, Geller et al. are skeptical of prediction claims over any period shorter than 30 years. A widely publicized TIP for an M 6.4 quake in Southern California in 2004 was not fulfilled, nor two other lesser known TIPs. A deep study of the RTP method in 2008 found that out of some twenty alarms only two could be considered hits (and one of those had a 60% chance of happening anyway). It concluded that "RTP is not significantly different from a naïve method of guessing based on the historical rates [of] seismicity."

Accelerating moment release (AMR, "moment" being a measurement of seismic energy), also known as time-to-failure analysis, or accelerating seismic moment release (ASMR), is based on observations that foreshock activity prior to a major earthquake not only increased, but increased at an exponential rate. In other words, a plot of the cumulative number of foreshocks gets steeper just before the main shock.

Following formulation by Bowman et al. (1998) into a testable hypothesis, and a number of positive reports, AMR seemed promising despite several problems. Known issues included not being detected for all locations and events, and the difficulty of projecting an accurate occurrence time when the tail end of the curve gets steep. But rigorous testing has shown that apparent AMR trends likely result from how data fitting is done, and failing to account for spatiotemporal clustering of earthquakes. The AMR trends are therefore statistically insignificant. Interest in AMR (as judged by the number of peer-reviewed papers) has fallen off since 2004.

The occurrence of foreshocks has long been thought to be the most promising avenue in predicting earthquakes. A foreshock is a smaller earthquake that can strike minutes or days before a larger one. Because the rupture process for the earthquakes is still not completely clear, foreshock occurrence may give clues into an earthquake-triggering process. In the Non-Critical Precursory Accelerating Seismicity Theory (N-C PAST), foreshocks happen because of the constant buildup of pressure along the fault lines. This theory is given weight due to seismic measurements. This had led to the conclusion for some scientists that foreshocks are a precursor to a larger event, and should be further studied and considered in earthquake prediction.

Notable Predictions

These are predictions, or claims of predictions, that are notable either scientifically or because of public notoriety, and claim a scientific or quasi-scientific basis. As many predictions are held confidentially, or published in obscure locations, and become notable only when they are claimed, there may be a selection bias in that hits get more attention than misses. The predictions listed here are discussed in Hough's book and Geller's paper.

1975: Haicheng, China

The M 7.3 1975 Haicheng earthquake is the most widely cited "success" of earthquake prediction. Study of seismic activity in the region led the Chinese authorities to issue a medium-term prediction in June 1974. The political authorities therefore ordered various measures taken, including enforced evacuation of homes, construction of "simple outdoor structures", and showing of movies out-of-doors. The quake, striking at 19:36, was powerful enough to destroy or badly damage about half of the homes. However, the "effective preventative measures taken" were said to have kept the death toll under 300 in an area with population of about 1.6 million, where otherwise tens of thousands of fatalities might have been expected.

However, although a major earthquake occurred, there has been some skepticism about the narrative of measures taken on the basis of a timely prediction. This event occurred during the Cultural Revolution, when "belief in earthquake prediction was made an element of ideological orthodoxy that distinguished the true party liners from right wing deviationists". Recordkeeping was disordered, making it difficult to verify details, including whether there was any ordered evacuation. The method used for either the medium-term or short-term predictions (other than "Chairman Mao's revolutionary line") has not been specified. The evacuation may have been spontaneous, following the strong (M 4.7) foreshock that occurred the day before.

A 2006 study that had access to an extensive range of records found that the predictions were flawed. "In particular, there was no official short-term prediction, although such a prediction was made by individual scientists." Also: "it was the foreshocks alone that triggered the final decisions of warning and evacuation". They estimated that 2,041 lives were lost. That more did not die was attributed to a number of fortuitous circumstances, including earthquake education in the previous months (prompted by elevated seismic activity), local initiative, timing (occurring when people were neither working nor asleep), and local style of construction. The authors conclude that, while unsatisfactory as a prediction, "it was an attempt to predict a major earthquake that for the first time did not end up with practical failure."

1985–1993: Parkfield, U.S. (Bakun-Lindh)

The "Parkfield earthquake prediction experiment" was the most heralded scientific earthquake prediction ever. It was based on an observation that the Parkfield segment of the San Andreas Fault breaks regularly with a moderate earthquake of about M 6 every several decades: 1857, 1881, 1901, 1922, 1934, and 1966. More particularly, Bakun & Lindh (1985) pointed out that, if the 1934 quake is excluded, these occur every 22 years, ±4.3 years. Counting from 1966, they predicted a 95% chance that the next earthquake would hit around 1988, or 1993 at the latest. The National Earthquake Prediction Evaluation Council (NEPEC) evaluated this, and concurred. The U.S. Geological Survey and the State of California therefore established one of the "most sophisticated and densest nets of monitoring instruments in the world", in part to identify any precursors when the quake came. Confidence was high enough that detailed plans were made for alerting emergency authorities if there were signs an earthquake was imminent. In the words of the Economist: "never has an ambush been more carefully laid for such an event."

1993 came, and passed, without fulfillment. Eventually there was an M 6.0 earthquake on the Parkfield segment of the fault, on 28 September 2004, but without forewarning or obvious precursors. While the *experiment* in catching an earthquake is considered by many scientists to have been successful, the *prediction* was unsuccessful in that the eventual event was a decade late.

1983–1995: Greece (VAN)

In 1981, the "VAN" group, headed by Panayiotis Varotsos, said that they found a relationship between earthquakes and 'seismic electric signals' (SES). In 1984 they claimed there was a "one-to-one correspondence" between SES and earthquakes", – that is, that *"every sizable EQ is preceded by an SES and inversely every SES is always followed by an EQ* the magnitude and the epicenter of which can be reliably predicted" – the SES appearing between six and 115 hours before the earthquake.. As validation they presented a table of 23 earthquakes from 19 January 1983 to 19 September 1983, of which they claimed to have successfully predicted 18 earthquakes. Their report was "saluted by some as a major breakthrough", but was greeted by a "wave of generalized skepticism" among seismologists.

Other lists followed, such as their 1991 claim of predicting six out of seven earthquakes with $M_s \geq 5.5$ in the period of 1 April 1987 through 10 August 1989, or five out of seven earthquakes with $M_s \geq 5.3$ in the overlapping period of 15 May 1988 to 10 August 1989, In 1996 they published a "Summary of all Predictions issued from January 1st, 1987 to June 15, 1995", amounting to 94 predictions. Matching this against a list of "All earthquakes with M_s(ATH)" and within geographical bounds including most of Greece they come up with a list of 14 earthquakes they should have predicted. Here they claim ten successes, for a success rate of 70%, but also a false alarm rate of 89%.

The VAN predictions have been criticized on various grounds, including being geophysically implausible, "vague and ambiguous", failing to satisfy prediction criteria, and retroactive adjustment of parameters. A critical review of 14 cases where VAN claimed 10 successes showed only one case where an earthquake occurred within the prediction parameters. The VAN predictions not only fail to do better than chance, but show "a much better association with the events which occurred before them", according to Mulargia and Gasperini. Other early reviews found that the VAN results, when evaluated by definite parameters, were statistically significant. Both positive and negative views on VAN predictions from this period were summarized in the 1996 book "A Critical Review of VAN" edited by Sir James Lighthill and in a debate issue presented by the journal Geophysical Research Letters that was focused on the statistical significance of the VAN method. VAN had the opportunity to reply to their critics in those review publications. In 2011, the ICEF reviewed the 1996 debate, and concluded that the optimistic SES prediction capability claimed by VAN could not be validated.

A crucial issue is the large and often indeterminate parameters of the predictions, such that some critics say these are not predictions, and should not be recognized as such. Much of the controversy with VAN arises from this failure to adequately specify these parameters. Some of their telegrams include predictions of two distinct earthquake events, such as (typically) one earthquake predicted at 300 km "N.W" of Athens, and another at 240 km "W", "with magnitues [sic] 5,3 and 5,8", with no time limit.

VAN has disputed the 'pessimistic' conclusions of their critics, but the critics have not relented. It was suggested that VAN failed to account for clustering of earthquakes, or that they interpreted their data differently during periods of greater seismic activity.

VAN has been criticized on several occasions for causing public panic and widespread unrest. This has been exacerbated by the broadness of their predictions, which cover large areas of Greece (up to 240 kilometers across, and often pairs of areas), much larger than the areas actually affected by earthquakes of the magnitudes predicted (usually several tens of kilometers across). Magnitudes are similarly broad: a predicted magnitude of "6.0" represents a range from a benign magnitude 5.3 to a broadly destructive 6.7. Coupled with indeterminate time windows of a month or more, such predictions "cannot be practically utilized" to determine an appropriate level of preparedness, whether to curtail usual societal functioning, or even to issue public warnings.

2008: Greece (VAN)

After 2006, VAN claim that all alarms related to SES activity have been made public by posting at arxiv.org. Such SES activity is evaluated using a new method they call 'natural time'. One such report was posted on Feb. 1, 2008, two weeks before the strongest earthquake in Greece during the period 1983-2011. This earthquake occurred on February 14, 2008, with magnitude (Mw) 6.9. VAN's report was also described in an article in the newspaper Ethnos on Feb. 10, 2008. However, Gerassimos Papadopolous complained that the VAN reports were confusing and ambiguous, and that "none of the claims for successful VAN predictions is justified."

1989: Loma Prieta, U.S.

The 1989 Loma Prieta earthquake (epicenter in the Santa Cruz Mountains northwest of San Juan Bautista, California) caused significant damage in the San Francisco Bay Area of California. The U.S. Geological Survey (USGS) reportedly claimed, twelve hours *after* the event, that it had "forecast" this earthquake in a report the previous year. USGS staff subsequently claimed this quake had been "anticipated"; various other claims of prediction have also been made.

Harris (1998) reviewed 18 papers (with 26 forecasts) dating from 1910 "that variously offer or relate to scientific forecasts of the 1989 Loma Prieta earthquake." (In this case no distinction is made between a forecast, which is limited to a probabilistic estimate of an earthquake happening over some time period, and a more specific prediction.) None of these forecasts can be rigorously tested due to lack of specificity, and where a forecast does bracket the correct time and location, the window was so broad (e.g., covering the greater part of California for five years) as to lose any value as a prediction. Predictions that came close (but given a probability of only 30%) had ten- or twenty-year windows.

One debated prediction came from the M8 algorithm used by Keilis-Borok and associates in four forecasts. The first of these forecasts missed both magnitude (M 7.5) and time (a five-year window from 1 January 1984, to 31 December 1988). They did get the location, by including most of California and half of Nevada. A subsequent revision, presented to the NEPEC, extended the time window to 1 July 1992, and reduced the location to only central California; the magnitude remained the same. A figure they presented had two more revisions, for M ≥ 7.0 quakes in central California. The five-year time window for one ended in July 1989, and so missed the Loma Prieta event; the second revision extended to 1990, and so included Loma Prieta.

When discussing success or failure of prediction for the Loma Prieta earthquake, some scientists argue that it did not occur on the San Andreas fault (the focus of most of the forecasts), and involved dip-slip (vertical) movement rather than strike-slip (horizontal) movement, and so was not predicted. Other scientists argue that it did occur in the San Andreas fault *zone*, and released much of the strain accumulated since the 1906 San Francisco earthquake; therefore several of the forecasts were correct. Hough states that "most seismologists" do not believe this quake was *predicted* "per se". In a strict sense there were no predictions, only forecasts, which were only partially successful.

Iben Browning claimed to have predicted the Loma Prieta event, but (as will be seen in the next section) this claim has been rejected.

1990: New Madrid, U.S. (Browning)

Dr. Iben Browning (a scientist with a Ph.D. degree in zoology and training as a biophysicist, but no experience in geology, geophysics, or seismology) was an "independent business consultant" who forecast long-term climate trends for businesses. He supported the idea (scientifically unproven) that volcanoes and earthquakes are more likely to be triggered when the tidal force of the sun and the moon coincide to exert maximum stress on the earth's crust (syzygy). Having calculated when these tidal forces maximize, Browning then "projected" what areas were most at risk for a large earthquake. An area he mentioned frequently was the New Madrid Seismic Zone at the southeast corner of the state of Missouri, the site of three very large earthquakes in 1811–12, which he coupled with the date of 3 December 1990.

Browning's reputation and perceived credibility were boosted when he claimed in various promotional flyers and advertisements to have predicted (among various other events) the Loma Prieta earthquake of 17 October 1989. The National Earthquake Prediction Evaluation Council (NEPEC) formed an Ad Hoc Working Group (AHWG) to evaluate Browning's prediction. Its report (issued 18 October 1990) specifically rejected the claim of a successful prediction of the Loma Prieta earthquake. A transcript of his talk in San Francisco on 10 October showed he had said: "there will probably be several earthquakes around the world, Richter 6+, and there may be a volcano or two" – which, on a global scale, is about average for a week – with no mention of any earthquake in California.

Though the AHWG report disproved both Browning's claims of prior success and the basis of his "projection", it made little impact after a year of continued claims of a successful prediction. Browning's prediction received the support of geophysicist David Stewart, and the tacit endorsement of many public authorities in their preparations for a major disaster, all of which was amplified by massive exposure in the news media. Nothing happened on 3 December, and Browning died of a heart attack seven months later.

2004 & 2005: Southern California, U.S. (Keilis-Borok)

The M8 algorithm (developed under the leadership of Dr. Vladimir Keilis-Borok at UCLA) gained respect by the apparently successful predictions of the 2003 San Simeon and Hokkaido earthquakes. Great interest was therefore generated by the prediction in early 2004 of a M ≥ 6.4 earthquake to occur somewhere within an area of southern California of approximately 12,000 sq. miles, on or before 5 September

2004. In evaluating this prediction the California Earthquake Prediction Evaluation Council (CEPEC) noted that this method had not yet made enough predictions for statistical validation, and was sensitive to input assumptions. It therefore concluded that no "special public policy actions" were warranted, though it reminded all Californians "of the significant seismic hazards throughout the state." The predicted earthquake did not occur.

A very similar prediction was made for an earthquake on or before 14 August 2005, in approximately the same area of southern California. The CEPEC's evaluation and recommendation were essentially the same, this time noting that the previous prediction and two others had not been fulfilled. This prediction also failed.

2009: L'Aquila, Italy (Giuliani)

At 03:32 on 6 April 2009, the Abruzzo region of central Italy was rocked by a magnitude M 6.3 earthquake. In the city of L'Aquila and surrounding area around 60,000 buildings collapsed or were seriously damaged, resulting in 308 deaths and 67,500 people left homeless. Around the same time, it was reported that Giampaolo Giuliani had predicted the earthquake, had tried to warn the public, but had been muzzled by the Italian government.

Giampaolo Giuliani was a laboratory technician at the Laboratori Nazionali del Gran Sasso. As a hobby he had for some years been monitoring radon using instruments he had designed and built. Prior to the L'Aquila earthquake he was unknown to the scientific community, and had not published any scientific work. He had been interviewed on 24 March by an Italian-language blog, *Donne Democratiche*, about a swarm of low-level earthquakes in the Abruzzo region that had started the previous December. He said that this swarm was normal and would diminish by the end of March. On 30 March, L'Aquila was struck by a magnitude 4.0 temblor, the largest to date.

On 27 March Giuliani warned the mayor of L'Aquila there could be an earthquake within 24 hours, and an earthquake M~2.3 occurred. On 29 March he made a second prediction. He telephoned the mayor of the town of Sulmona, about 55 kilometers southeast of L'Aquila, to expect a "damaging" – or even "catastrophic" – earthquake within 6 to 24 hours. Loudspeaker vans were used to warn the inhabitants of Sulmona to evacuate, with consequential panic. No quake ensued and Giuliano was cited for inciting public alarm and enjoined from making future public predictions.

After the L'Aquila event Giuliani claimed that he had found alarming rises in radon levels just hours before. He said he had warned relatives, friends and colleagues on the evening before the earthquake hit. He was subsequently interviewed by the International Commission on Earthquake Forecasting for Civil Protection, which found that Giuliani had not transmitted a valid prediction of the mainshock to the civil authorities before its occurrence.

VAN Method

The VAN method – named after P. Varotsos, K. Alexopoulos and K. Nomicos, authors of the 1981 papers describing it – measures low frequency electric signals, termed "seismic electric signals" (SES), by which Varotsos and several colleagues claimed to have successfully predicted earthquakes in Greece. Both the method itself and the manner by which successful predictions were claimed have been severely criticized and debated by VAN, but the critics have not retracted their views. In 2001 the estimation of the time window of the forthcoming earthquake was changed to use a new analysis, termed "natural time". Research related to the VAN method is currently carried out at the Solid Earth Physics Institute, University of Athens, Greece.

Description of the VAN Method

Prediction of earthquakes with this method is based on the detection, recording and evaluation of seismic electric signals or SES. These electrical signals have a fundamental frequency component of 1 Hz or less and an amplitude the logarithm of which scales with the magnitude of the earthquake. According to VAN proponents, SES are emitted by rocks under stresses caused by plate-tectonic forces. There are three types of reported electric signal:

- Electric signals that occur shortly before a major earthquake. Signals of this type were recorded 6.5 hours before the 1995 Kobe earthquake in Japan, for example.

- Electric signals that occur some time before a major earthquake.

- A gradual variation in the Earth's electric field some time before an earthquake.

Several hypotheses have been proposed to explain SES:

- Stress-related phenomena: Seismic electric signals are perhaps attributed to the piezoelectric behaviour of some minerals, especially quartz, or to effects related to the behavior of crystallographic defects under stress or strain. Series of SES, termed SES activities (which are recorded before major earthquakes), may appear a few weeks to a few months before an earthquake when the mechanical stress reaches a critical value. The generation of electric signals by minerals under high stress leading to fracture has been confirmed with laboratory experiments.

- Thermoelectric phenomena: Alternately, Chinese researchers proposed a mechanism which relies on the thermoelectric effect in magnetite.

- Groundwater phenomena: Three mechanisms have been proposed relying on the presence of groundwater in generating SES. The electrokinetic effect is associated with the motion of groundwater during a change in pore pressure. The seismic dynamo effect is associated with the motion of ions in groundwater relative to the geomagnetic field as a seismic wave creates displacement. Circular polarization would be characteristic of the seismic dynamo effect, and this has been observed both for artificial and natural seismic events. A radon ionization effect, caused by radon release and then subsequent ionization of material in groundwater, may also be active. The main isotope of radon is radioactive with a half-life of 3.9 days, and the nuclear decay of radon is known to have an ionizing effect on air. Many publications have reported increased radon concentration in the vicinity of some active tectonic faults a few weeks prior to strong seismic events. However, a strong correlation between radon anomalies and seismic events has not been demonstrated.

While the electrokinetic effect may be consistent with signal detection tens or hundreds of kilometers away, the other mechanisms require a second mechanism to account for propagation:

- Signal transmission along faults: In one model, seismic electric signals propagate with relatively low attenuation along tectonic faults, due to the increased electrical conductivity caused either by the intrusion of ground water into the fault zone(s) or by the ionic characteristics of the minerals.

- Rock circuit: In the defect model, the presence of charge carriers and holes can be modeled as making an extensive circuit.

Seismic electric signals are detected at stations which consist of pairs of electrodes (oriented NS and EW) inserted into the ground, with amplifiers and filters. The signals are then transmitted to the VAN scientists in Athens where they are recorded and evaluated. Currently the VAN team operates 9 stations, while in the past (until 1989) they could afford up to 17.

The VAN team claimed that they were able to predict earthquakes of magnitude larger than 5, with an uncertainty of 0.7 units of magnitude, within a radius of 100 km, and in time window ranging from several hours to a few weeks. Several papers confirmed this success rate, leading to statistically significant conclusion. For example, there were eight M ≥ 5.5 earthquakes in Greece from January 1, 1984 through September 10, 1995, and the VAN network forecast six of these.

The VAN method has also been used in Japan, but in early attempts success comparable to that achieved in Greece was "difficult" to attain. A preliminary investigation of seismic electric signals in France led to encouraging results.

Earthquake Prediction using "Natural Time" Analysis

Since 2001 the VAN team has attempted to improve the accuracy of the estimation of the time of the forthcoming earthquake. To that end, they introduced the concept of *natural time*, a time series analysis technique which puts weight on a process based on the ordering of events. Two terms characterize each event, the "natural time" χ, and the energy Q. χ is defined as k/N, where k is an integer (the k-th event) and N is the total number of events in the time sequence of data. A related term, p_k, is the ratio Q_k / Q_{total}, which describes the fractional energy released. They introduce a critical term κ, the "variance in natural time", which puts extra weight on the energy term p_k:

$$\kappa = \sum_{k=1}^{N} p_k (\chi_k)^2 - \left(\sum_{k=1}^{N} p_k \chi_k \right)^2$$

where $\chi_k = k / N$ and $p_k = \dfrac{Q_k}{\sum_{n=1}^{N} Q_n}$ which gives

$$\kappa = \sum_{k=1}^{N} p_k (k / N)^2 - \left(\sum_{k=1}^{N} p_k (k / N) \right)^2$$

Their current method detects SES as valid when $\kappa = 0.070$. Once the SES are deemed valid, a second analysis is started in which the subsequent seismic (rather than electric) events are noted, and the region is divided up as a Venn diagram with at least two seismic events per overlapping rectangle. When the distribution of κ for the rectangular regions has its maximum at $\kappa = 0.070$, a critical seismic event is imminent, i.e. it will occur in a few days to one week or so, and a report is issued.

Results

The VAN team claim that out of seven mainshocks with magnitude Mw>=6.0 from 2001 through 2010 in the region of latitude N 36° to N 41° and longitude E 19° to E 27°, all but one could be classified with relevant SES activity identified and reported in advance through natural time analysis. Additionally, they assert that the occurrence time of four of these mainshocks with magnitude Mw>=6.4 were identified to within "a narrow range, a few days to around one week or so." These reports are inserted in papers housed in arXiv, and new reports are made and uploaded there. For example, a report preceding the strongest earthquake in Greece during the period 1983-2011, which occurred on February 14, 2008, with magnitude (Mw) 6.9, was publicized in arXiv almost two weeks before, on February 1, 2008. A description of the updated VAN method was collected in a book published by Springer in 2011, titled "Natural Time Analysis: The New View of Time."

Natural time analysis also claims that the physical connection of SES activities with earthquakes is as follows: Taking the view that the earthquake occurrence is a phase-change (critical phenomenon), where the new phase is the mainshock occurrence, the above-mentioned variance term κ is the corresponding order parameter. The κ value calculated for a window comprising a number of seismic events comparable to the average number of earthquakes occurring within a few months, fluctuates when the window is sliding through a seismic catalogue. The VAN team claims that these κ fluctuations exhibit a minimum a few months before a mainshock occurrence and in addition this minimum occurs simultaneously with the initiation of the corresponding SES activity, and that this is the first time in the literature that such a simultaneous appearance of two precursory phenomena in independent datasets of different geophysical observables (electrical measurements, seismicity) has been observed. Furthermore, the VAN team claims that their natural time analysis of the seismic catalogue of Japan during the period from January 1, 1984 until the occurrence of the magnitude 9.0 Tohoku earthquake on March 11, 2011, revealed that such clear minima of the κ fluctuations appeared before all major earthquakes with magnitude 7.6 or larger. The deepest of these minima was said to occur on January 5, 2011, i.e., almost two months before the Tohoku earthquake occurrence. Finally, by dividing the Japanese region into small areas, the VAN team states that some small areas show minimum of the κ fluctuations almost simultaneously with the large area covering the whole Japan and such small areas clustered within a few hundred kilometers from the actual epicenter of the impending major earthquake.

Criticisms of VAN

Currently, the major criticism of the VAN method is that results have not yet been replicated by scientists outside specific research groups in Greece and Japan. Testing of the method needs to be done by scientists unrelated to these two groups. Independent verification is a standard protocol in science.

Historically, the usefulness of the VAN method for prediction of earthquakes had been a matter of debate. Both positive and negative criticism on an older conception of the VAN method is summarized in the 1996 book "A Critical Review of VAN", edited by Sir James Lighthill. A critical review of the statistical methodology was published by Y. Y. Kagan of UCLA in 1997. Note that these criticisms predate the time series analysis methods introduced by the VAN group in 2001. The main points of the criticism were:

- Predictive success: VAN has claimed to have observed at a recording station in Athens a perfect record of a one-to-one correlation between SESs and earthquake of magnitude ≥ 2.9 which occurred 7 hours later in all of Greece. However, it was later shown that the list of earthquake used for the correlation was false. Although VAN stated in their article that the list of earthquakes was that of the Bulletin of the National Observatory of Athens (NOA), it was found that 37% of the earthquakes actually listed in the bulletin, including the largest one, were not in the list used by VAN for issuing their claim. In addition, 40% of the earthquake which VAN claimed had occurred were not in the NOA bulletin. Examining the probability of chance correlation of 22 claims of successful predictions by VAN of M > 4.0 from January 1, 1987 through November 30, 1989 it was found that 74% were false, 9% correlated by chance, and for 14% the correlation was uncertain. No single event correlated at a probability greater than 85%, whereas the level required in statistics for accepting a hypothesis test as positive would more commonly be 95%.

- Proposed SES propagation mechanism: An analysis of the propagation properties of SES in the Earth's crust showed that it is impossible that signals with the amplitude reported by VAN could have been generated by small earthquakes and transmitted over the several hundred kilometers between the epicenter and the receiving station. In effect, if the mechanism is based on piezoelectricity or electrical charging of crystal deformations with the signal traveling along faults, then none of the earthquakes which VAN claimed were preceded by SES generated an SES themselves. There is also some doubt that the phenomena of rock physics seen in certain laboratory conditions can be assumed to take place in the Earth's seismogenic crust.

- Electromagnetic compatibility issues: VAN's publications are further weakened by failure to address the problem of eliminating the many and strong sources of change in the magneto-electric field measured by them, such as telluric currents from weather, and electromagnetic interference (EMI) from man-made signals. One critical paper clearly correlates an SES used by the VAN group with digital radio transmissions made from a military base.

- Scientific reporting: Most importantly, the method is hindered by a lack of statistical testing of the validity of the VAN hypothesis because the researchers keep changing the parameters of the hypothesis (the moving the goalposts) technique).

- Public policy: Finally, one requirement for any earthquake prediction method is that, in order for any prediction to be useful, it must predict a forthcoming earthquake within a reasonable time-frame, epicenter and magnitude. If the prediction is too vague, no feasible decision (such as to evacuate the population of a certain area for a given period of time) can be made. In practice, the VAN group issued a series of telegrams in the 1980s, warning of impending earthquakes that did not occur, or did not occur within the parameters listed in the telegrams. During the same time frame, the technique also missed major earthquakes. These inaccurate predictions from the early VAN method led to public criticism and the cost associated with false alarms generated ill will. Major opponents of VAN were the Greek seismologists Vassilis Papazachos and G. Stavrakakis. The debate between Papazachos and the VAN team has repeatedly caused public attention in their home country Greece and has been extensively discussed in the Greek media.

References

- Ohnaka, M. (2013). The Physics of Rock Failure and Earthquakes. Cambridge University Press. p. 148. ISBN 9781107355330.

- Van Riper, A. Bowdoin (2002). Science in popular culture: a reference guide. Westport: Greenwood Press. p. 60. ISBN 0-313-31822-0.

- Gates, A.; Ritchie, D. (2006). Encyclopedia of Earthquakes and Volcanoes. Infobase Publishing. p. 89. ISBN 978-0-8160-6302-4. Retrieved 29 November 2010.

- Kayal, J.R. (2008). Microearthquake seismology and seismotectonics of South Asia. Springer. p. 15. ISBN 978-1-4020-8179-8. Retrieved 29 November 2010.

- Lazaridou-Varotsos, M. (2013), Earthquake Prediction by Seismic Electric Signals: The Success of the VAN method over thirty years, Springer-Praxis, doi:10.1007/978-3-642-24406-3, ISBN 978-3-642-24405-6

- Lighthill, Sir James, ed. (1996), A Critical Review of VAN – Earthquake Prediction from Seismic Electrical Signals, London, UK: World Scientific Publishing Co Pte Ltd, ISBN 978-981-02-2670-1

- Mulargia, F.; Geller, R. (2003), Earthquake Science and Seismic Risk Reduction, Springer, ISBN 978-1402017773

- Varotsos, P.; Sarlis, N.; Skordas, E. (2011), Natural time analysis : the new view of time ; Precursory seismic electric signals, earthquakes and other complex time series, Springer Praxis, ISBN 364216448-X

- Record tying Oklahoma earthquake felt as far away as Arizona, Associated Press, Ken Miller, September 3, 2016. Retrieved 3 September 2016.

- USGS calls for shut down of wells, governor declares emergency in wake of 5.6 quake in Oklahoma, Enid News & Eagle, Sally Asher & Violet Hassler, September 3, 2016. Retrieved 4 September 2016.

- Battaglia, Steven M.; Changnon, David (2016-01-02). "Frost Quakes: Forecasting the Unanticipated Clatter". Weatherwise. 69 (1): 20–27. doi:10.1080/00431672.2015.1109984. ISSN 0043-1672.

- Manninen, Tuomas (22 January 2016). "Mystisiä pamahduksia eri puolilla Suomea, kivitalon seinä halkesi – mistä on kyse?". Ilta-Sanomat (in Finnish). Retrieved 22 January 2016.

- Keller, G. Randy; Holland, Austin A. (March 22, 2013). Statement about the cause of 2011 Prague Earthquake Sequence (PDF). Oklahoma Geological Survey (Report). Retrieved April 30, 2015.

- Pérez-Peña, Richard (April 23, 2015). "U.S. Maps Pinpoint Earthquakes Linked to Quest for Oil and Gas". New York Times. Retrieved November 8, 2015.

- Andrews, Richard D.; Holland, Austin A. (April 21, 2015). Statement on Oklahoma Seismicity (PDF). Oklahoma Geological Survey (Report). University of Oklahoma. Retrieved April 30, 2015.

- Walsh, F.R.; Zoback, M.D. (2015). "Oklahoma's recent earthquakes and saltwater disposal". Science Advances. 1 (5): e1500195. doi:10.1126/sciadv.1500195.

- Weingarten, Matthew; Ge, Shemin; Godt, J.W.; Bekins, B.A.; Rubinstein, J.L. (2015). "High-rate injection is associated with the increase in U.S. mid-continent seismicity". Science. 348 (6241): 1336–1340. doi:10.1126/science.aab1345.

- Brown E. (May 11, 2012). "Quake study offers new clues on a California fault's mystery". Los Angelel Times. Retrieved 30 March 2015.

- Ellsworth, W.L. (2013). "Injection-induced earthquakes". Science. 341 (6142): 7. doi:10.1126/science.1225942.

- Geographic.org. "Magnitude 8.0 - SANTA CRUZ ISLANDS Earthquake Details". Gobal Earthquake Epicenters with Maps. Retrieved 2013-03-13.

- Henry Fountain (March 28, 2013). "Study Links 2011 Quake to Technique at Oil Wells". The New

York Times. Retrieved March 29, 2013.

- Nicole Mortillaro (25 December 2013). "Mysterious Christmas Eve 'boom' heard and felt around GTA". Global News.

- The Canadian Press (23 December 2013). "Toronto ice storm 2013: Photos show city looking like crime scene with taped-off downed branches". National Post.

- Kagan & Jackson 1991, pp. 21,420; Stein, Friedrich & Newman 2005; Jackson & Kagan 2006; Tiampo & Shcherbakov 2012, §2.2, and references there; Kagan, Jackson & Geller 2012.

Earthquake Preparedness and Mitigation

6

The extent of damage earthquakes can cause is very extreme. Governments take steps in order to analyze and then to take precautionary steps regarding them. Some of these are earthquake preparedness, earthquake warning system, seismic analysis and seismic retrofit. Earthquakes can best be understood in confluence with the major topics listed in the following chapter.

Earthquake Preparedness

Earthquake preparedness is a set of measures taken at the individual, organisational and societal level to minimise the effects of an earthquake. Preparedness measures can range from securing heavy objects, structural modifications and storing supplies, to having insurance, an emergency kit, and evacuation plans.

Preparedness Measures

Preparedness can consist of survival measures, preparation that will improve survival in the event of an earthquake, or mitigating measures, that seek to minimise the effect of an earthquake. Common survival measures include storing food and water for an emergency, and educating individuals what to do during an earthquake. Mitigating measures can include firmly securing large items of furniture (such as bookcases and large cabinets), TV and computer screens that may otherwise fall over in an earthquake. Likewise, avoiding storing items above beds or sofas reduces the chance of objects falling on individuals.

Planning for a related tsunami, tsunami preparedness, can also be part of earthquake preparedness.

Building Design and Retrofitting

Building codes in earthquake prone areas may have specific requirements designed to increase new buildings' resistance to earthquakes. Older buildings and homes that are not up to code may be modified to increase their resistance. Modification and earthquake resistant design are also employed in elevated freeways and bridges.

Codes are not designed to make buildings earthquake proof in the sense of them suffering zero damage. The goal of most building designs is to reduce earthquake damage to a building such that it protects the lives of occupants and thus tolerance of some limited damage is accepted and considered a necessary tradeoff. A supplement or precursor to retrofitting can be the implementation of earthquake proof furniture.

Earthquake modification techniques and modern building codes are designed to prevent total destruction of buildings for earthquakes of no greater than 8.5 on the Richter Scale. Although the Richter Scale is referenced, the localized shaking intensity is one of the largest factors to be considered in building resiliency.

Types of Preparedness

The basic theme behind preparedness is to be ready for an earthquake. Preparedness starts with an individual's everyday life and involves items and training that would be useful in an earthquake. Preparedness continues on a continuum from individual preparedness through family preparedness, community preparedness and then business, non-profit and governmental preparedness. Some organisations blend these various levels. Business continuity planning encourages businesses to have a Disaster Recovery Plan. The US FEMA breaks down preparedness generally into a pyramid, with citizens on the foundational bottom, on top of which rests local government, state government and federal government in that order.

Non-perishable food in cabinet

Children may present particular issues and some planning and resources are directly focused on supporting them. The US FEMA has advice noting that "Disasters can leave children feeling frightened, confused, and insecure" whether a child has experienced it first hand, had it happen to a friend or simply seen it on television. People with disabilities or other special needs may have special emergency preparation needs. FEMA's suggestions for people with disabilities include having copies of prescriptions, charging devices for medical devices such as motorized wheel chairs and a week's supply of medication readily available. Preparedness can also cover pets.

Preparedness can also encompass psychological preparedness: resources are designed to support both community members affected by a disaster and the disaster workers serving them.

A multi-hazard approach, where communities are prepared for several hazards, are more resilient than single hazard approaches and have been gaining popularity.

Long term power outages can cause damage beyond the original disaster that can be mitigated with emergency generators or other power sources to provide an emergency power system. The United States Department of Energy states: "homeowners, business owners, and local leaders may have to take an active role in dealing with energy disruptions on their own." Major institutions like hospitals, military bases and educational institutions often have extensive backup power systems. Preparedness does not stop at home or at school. The United States Department of Health and Human Services addresses specific emergency preparedness issues hospitals may have to respond to, including maintaining a safe temperature, providing adequate electricity for life support systems and even carrying out evacuations under extreme circumstances. FEMA encourages all businesses to have businesses to have an emergency response plan and the Small Business Administration specifically advises small business owners to also focus emergency preparedness and provides a variety of different worksheets and resources.

Marked gas shuttoff

Given the explosive danger posed by natural gas leaks, Ready.gov states that "It is vital that all household members know how to shut off natural gas" and that property owners must ensure they have any special tools needed for their particular gas connections.

Ready.gov also notes that "It is wise to teach all responsible household members where and how to shut off the electricity," cautioning that individual circuits should be shut off before the main circuit. Ready.gov further states that "It is vital that all household members learn how to shut off the water at the main house valve" and cautions that the possibility that rusty valves might require replacement.

Achieving Preparedness

Levels of preparedness generally remain low, despite attempts to increase public awareness.

Various methods exist to promote disaster preparedness, but they are rarely well documented and their efficacy is rarely tested. Hands on training, drills and face-to-face interaction have proven more successful at changing behaviour. Digital methods have also been used, including for examples educational videogames.

Seismic Hazard

A seismic hazard is the probability that an earthquake will occur in a given geographic area, within a given window of time, and with ground motion intensity exceeding a given threshold. With a hazard thus estimated, risk can be assessed and included in such areas as building codes for standard buildings, designing larger buildings and infrastructure projects, land use planning and determining insurance rates. The seismic hazard studies also may generate two standard measures of anticipated ground motion, both confusingly abbreviated MCE; the simpler probabilistic Maximum Considered Earthquake (or Event), used in standard building codes, and the more detailed and deterministic Maximum Credible Earthquake incorporated in the design of larger buildings and civil infrastructure like dams or bridges. It is important to clarify which MCE is being discussed.

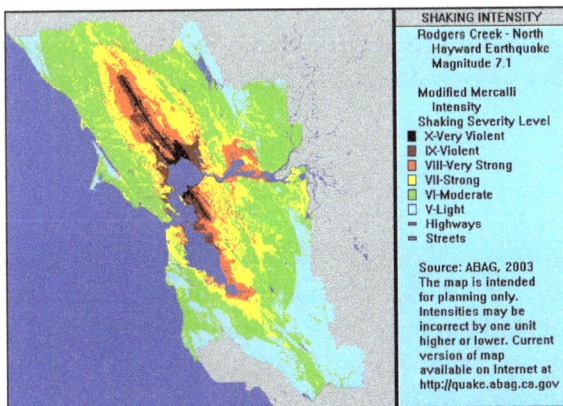

Surface motion map for a hypothetical earthquake on the northern portion of the Hayward Fault Zone and its presumed northern extension, the Rodgers Creek Fault Zone

Calculations for determining seismic hazard were first formulated by C. Allin Cornell in 1968 and, depending on their level of importance and use, can be quite complex. The regional geology and seismology setting is first examined for sources and patterns of earthquake occurrence, both in depth and at the surface from seismometer records; secondly, the impacts from these sources are assessed relative to local geologic rock and soil types, slope angle and groundwater conditions. Zones of similar potential earthquake shaking are thus determined and drawn on maps. The well known San Andreas Fault is illustrated as a long narrow elliptical zone of greater potential motion, like many areas along continental margins associated with the Pacific ring of fire. Zones of higher seismicity in the continental interior may be the site for intraplate earthquakes) and tend to be drawn as broad areas, based on historic records, like the 1812 New Madrid earthquake, since specific causative faults are generally not identified as earthquake sources.

Each zone is given properties associated with source potential: how many earthquakes per year, the maximum size of earthquakes (maximum magnitude), etc. Finally, the calculations require formulae that give the required hazard indicators for a given earthquake size and distance. For example, some districts prefer to use peak acceleration, others use peak velocity, and more sophisticated uses require response spectral ordinates.

The computer program then integrates over all the zones and produces probability curves for the key ground motion parameter. The final result gives a 'chance' of exceeding a given value over a specified amount of time. Standard building codes for homeowners might be concerned with a 1 in 500 years chance, while nuclear plants look at the 10,000 year time frame. A longer-term seismic history can be obtained through paleoseismology. The results may be in the form of a ground response spectrum for use in seismic analysis.

More elaborate variations on the theme also look at the soil conditions. Higher ground motions are likely to be experienced on a soft swamp compared to a hard rock site. The standard seismic hazard calculations become adjusted upwards when postulating characteristic earthquakes. Areas with high ground motion due to soil conditions are also often subject to soil failure due to liquefaction. Soil failure can also occur due to earthquake-induced landslides in steep terrain. Large area landsliding can also occur on rather gentle slopes as was seen in the Good Friday earthquake in Anchorage, Alaska, March 28, 1964.

MCEs

In a normal seismic hazard analyses intended for the public, that of a "maximum considered earthquake", or "maximum considered event" (MCE) for a specific area, is an earthquake that is expected to occur once in approximately 2,500 years; that is, it has a 2-percent probability of being exceeded in 50 years. The term is used specifically for general building codes, which people commonly occupy; building codes in many localities will require non-essential buildings to be designed for "collapse prevention" in an MCE, so that the building remains standing - allowing for safety and escape of occupants - rather than full structural survival of the building.

A far more detailed and stringent MCE stands for "maximum credible earthquake", which is used in designing for skyscrapers and larger civil infrastructure, like dams, where structural failure could lead to other catastrophic consequences. These MCEs might require determining more than one specific earthquake event, depending on the variety of structures included.

US Seismic Hazard Maps

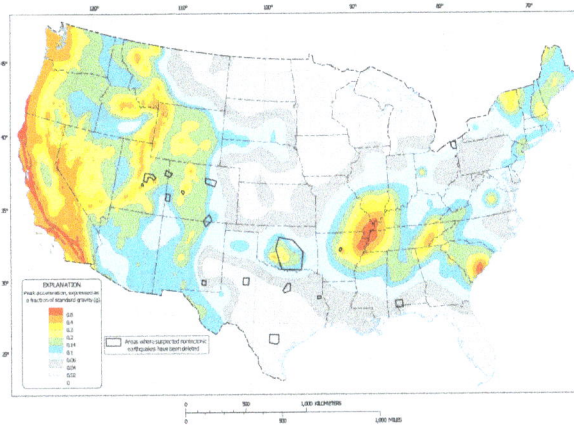

Two-percent probability of exceedance in 50 years map of peak ground acceleration

Map of peak ground acceleration with 2% probability of exceedance in 50 years

Some maps released by the USGS are shown with peak ground acceleration with a 10% probability of exceedance in 50 years, measured in Metre per second squared. For parts of the US, the National Seismic Hazard Mapping Project in 2008 resulted in seismic hazard maps showing peak acceleration (as a percentage of gravity) with a 2% probability of exceedance in 50 years.

Temblor, a company founded in 2014, offers a seismic hazard rank for the all of the conterminous US. This service is free and ad-free for the public. The hazard rank "is made for the likelihood of experiencing strong shaking (0.4g peak ground acceleration) in 30 years, based on the 2014 USGS NSHMP hazard model."

Earthquake Warning System

An earthquake warning system is a system of accelerometers, seismometers, communication, computers, and alarms that is devised for regional notification of a substantial earthquake while it is in progress. This is not the same as earthquake prediction, which is currently incapable of producing decisive event warnings.

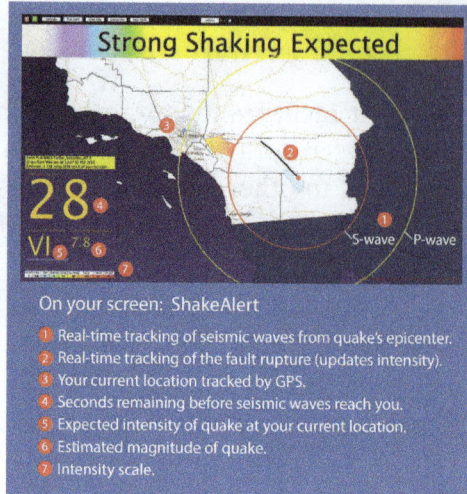

Example of early warning issued by ShakeAlert.

Time Lag and Wave Projection

An earthquake is caused by the release of stored elastic strain energy during rapid sliding along a fault. The sliding will start at some location and progress away from this hypo-center in each direction along the fault surface. The speed of the progression of this fault tear is slower than and distinct from the speed of the resultant pressure and shear waves, with the pressure wave traveling faster than the shear wave. The pressure wave will generate an abrupt shock while the shear waves can generate a periodic motion (at about 1 Hz) that is the most destructive in its effect upon structures, particularly buildings that have a similar resonant period, typically buildings around eight floors in height. These waves will be strongest at the ends of the slippage, and may project destructive waves well beyond the fault failure. The intensity of such remote effects are highly dependent upon local soils conditions within the region and these effects are considered in constructing a computer model of the region that determines appropriate responses to specific events.

Transit Safety

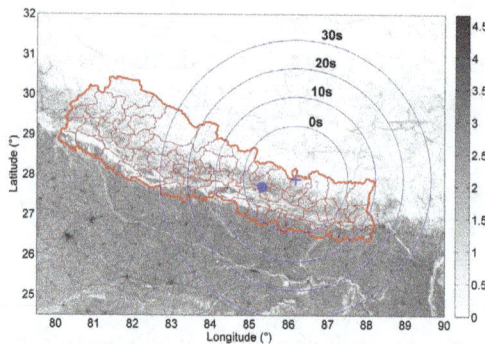

Warning time given by the earthquake warning system of the Earthquake Network project during the May 2015 Nepal earthquake. The cross marker depicts the earthquake epicenter while the dot marker shows the detection location.

Such systems are currently implemented to determine appropriate real-time response to an event in determining train operator response for urban rail systems such as BART (Bay Area Rapid Transit). The appropriate response will be highly dependent upon the warning time, the local right–of–way conditions, and the current speed of the train.

Deployment

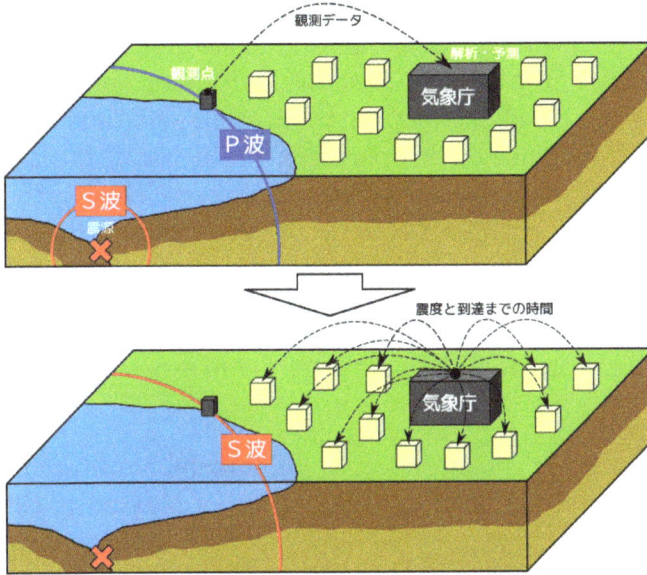

The Earthquake Early Warning in Japan: When P-waves are detected, the readings are analyzed immediately and the warning information is distributed to advanced users such as; broadcasting stations and mobile phone companies, before the arrival of S-waves (lower).

As of 2016, Japan is the only country with a comprehensive nationwide earthquake early warning system. Other countries and regions have limited deployment of earthquake warning systems, including Taiwan, Mexico (installed to issue alerts to Mexico City primarily), limited regions of Romania (the Basarab bridge in Bucharest) and parts of the United States. The earliest automated earthquake pre-detection systems were installed in the 1990s, for instance in California the Calistoga fire station's system which can automatically trigger a citywide siren to alert the entire area's residents. While many of these efforts are governmental, several private companies also manufacturer earthquake early warning systems to protect infrastructure such as elevators, gas lines and fire stations.

Japan's Earthquake Early Warning system was put to practical use in 2006. Its scheme to warn the general public was installed on October 1, 2007. It was modeled partly on the Urgent Earthquake Detection and Alarm System (UrEDAS) of Japan Railways, which was designed to enable automatic braking of bullet trains.

In 2009, an early warning system called ShakeAlarm was installed and commissioned in Vancouver British Columbia Canada. It was placed to protect a piece of critical transportation infrastructure called the George Massey Tunnel, which connects north and south banks of the Fraser River. In this application the system automatically closes the gates at the tunnel entrances if there is a dangerous seismic event inbound. The success and the reliability of the system was such that as of 2015 there have been several additional installations on the West coast of Canada and the United States, and there are more being planned.

In January 2013, Francesco Finazzi of the University of Bergamo started the Earthquake Network research project which aims at developing and maintaining a crowdsourced earthquake warning system based on smartphone networks. Smartphones are used to detect the ground shaking induced by an earthquake and a warning is issued as soon as an earthquake is detected. People living at a higher distance from the epicenter and the detection point may be alerted before they are reached by the damaging waves of the earthquake. People can take part in the project installing the Android application "Earthquake Network" which is also required to receive the alerts.

In December 2014, United States Congress approved a $5 million allocation as part of the Consolidated Appropriations Act, 2014 in order to expand funding for development of the system.

The USGS has investigated collaboration with the social networking site Twitter to allow for more rapid construction of ShakeMaps. As of May 2015, USGS began testing an earthquake early warning app. QuakeAlert was developed by Early Warning Labs in partnership USGS.

In July 2015, the USGS awarded $4 million in funding to the Berkeley Seismological Laboratory, the California Institute of Technology and the University of Washington, to turn the current ShakeAlert early warning prototype into a more robust system that could be used by "cities, industries, utilities and transportation networks in California, Oregon and Washington".

Seismic Analysis

First and second modes of building seismic response

Seismic analysis is a subset of structural analysis and is the calculation of the response of a building (or nonbuilding) structure to earthquakes. It is part of the process of structural design, earthquake engineering or structural assessment and retrofit in regions where earthquakes are prevalent.

As seen in the figure, a building has the potential to 'wave' back and forth during an earthquake (or even a severe wind storm). This is called the 'fundamental mode', and is the lowest frequency of building response. Most buildings, however, have higher modes of response, which are uniquely activated during earthquakes. The figure just shows the second mode, but there are higher 'shimmy' (abnormal vibration) modes. Nevertheless, the first and second modes tend to cause the most damage in most cases.

The earliest provisions for seismic resistance were the requirement to design for a lateral force equal to a proportion of the building weight (applied at each floor level). This approach was adopted in the appendix of the 1927 Uniform Building Code (UBC), which was used on the west coast of the United States. It later became clear that the dynamic properties of the structure affected the loads generated during an earthquake. In the Los Angeles County Building Code of 1943 a provision to vary the load based on the number of floor levels was adopted (based on research carried out at Caltech in collaboration with Stanford University and the U.S. Coast and Geodetic Survey, which started in 1937). The concept of "response spectra" was developed in the 1930s, but it wasn't until 1952 that a joint committee of the San Francisco Section of the ASCE and the Structural Engineers Association of Northern California (SEAONC) proposed using the building period (the inverse of the frequency) to determine lateral forces.

The University of California, Berkeley was an early base for computer-based seismic analysis of structures, led by Professor Ray Clough (who coined the term finite element). Students included Ed Wilson, who went on to write the program SAP in 1970, an early "Finite Element Analysis" program.

Earthquake engineering has developed a lot since the early days, and some of the more complex designs now use special earthquake protective elements either just in the foundation (base isolation) or distributed throughout the structure. Analyzing these types of structures requires specialized explicit finite element computer code, which divides time into very small slices and models the actual physics, much like common video games often have "physics engines". Very large and complex buildings can be modeled in this way (such as the Osaka International Convention Center).

Structural analysis methods can be divided into the following five categories.

Equivalent Static Analysis

This approach defines a series of forces acting on a building to represent the effect of earthquake ground motion, typically defined by a seismic design response spectrum. It assumes that the building responds in its fundamental mode. For this to be true, the building must be low-rise and must not twist significantly when the ground moves. The response is read from a design response spectrum, given the natural frequency of the building (either calculated or defined by the building code). The applicability of this method is extended in many building codes by applying factors to account for higher buildings with some higher modes, and for low levels of twisting. To account for effects due to "yielding" of the structure, many codes apply modification factors that reduce the design forces (e.g. force reduction factors).

Response Spectrum Analysis

This approach permits the multiple modes of response of a building to be taken into account (in the frequency domain). This is required in many building codes for all except very simple or very complex structures. The response of a structure can be defined as a combination of many special shapes (modes) that in a vibrating string correspond to the "harmonics". Computer analysis can be used to determine these modes for a structure. For each mode, a response is read from the design spectrum, based on the modal frequency and the modal mass, and they are then combined to provide an estimate of the total response of the structure. In this we have to calculate the magnitude of forces in all directions i.e. X, Y & Z and then see the effects on the building.. Combination methods include the following:

- absolute - peak values are added together

- square root of the sum of the squares (SRSS)

- complete quadratic combination (CQC) - a method that is an improvement on SRSS for closely spaced modes

The result of a response spectrum analysis using the response spectrum from a ground motion is typically different from that which would be calculated directly from a linear dynamic analysis using that ground motion directly, since phase information is lost in the process of generating the response spectrum.

In cases where structures are either too irregular, too tall or of significance to a community in disaster response, the response spectrum approach is no longer appropriate, and more complex analysis is often required, such as non-linear static analysis or dynamic analysis.

Linear Dynamic Analysis

Static procedures are appropriate when higher mode effects are not significant. This is generally true for short, regular buildings. Therefore, for tall buildings, buildings with torsional irregularities, or non-orthogonal systems, a dynamic procedure is required. In the linear dynamic procedure, the building is modelled as a multi-degree-of-freedom (MDOF) system with a linear elastic stiffness matrix and an equivalent viscous damping matrix.

The seismic input is modelled using either modal spectral analysis or time history analysis but in both cases, the corresponding internal forces and displacements are determined using linear elastic analysis. The advantage of these linear dynamic procedures with respect to linear static procedures is that higher modes can be considered. However, they are based on linear elastic response and hence the applicability decreases with increasing nonlinear behaviour, which is approximated by global force reduction factors.

In linear dynamic analysis, the response of the structure to ground motion is calculated in the time domain, and all phase information is therefore maintained. Only linear properties are assumed. The analytical method can use modal decomposition as a means of reducing the degrees of freedom in the analysis.

Nonlinear Static Analysis

In general, linear procedures are applicable when the structure is expected to remain nearly elastic for the level of ground motion or when the design results in nearly uniform distribution of nonlinear response throughout the structure. As the performance objective of the structure implies greater inelastic demands, the uncertainty with linear procedures increases to a point that requires a high level of conservatism in demand assumptions and acceptability criteria to avoid unintended performance. Therefore, procedures incorporating inelastic analysis can reduce the uncertainty and conservatism.

This approach is also known as "pushover" analysis. A pattern of forces is applied to a structural model that includes non-linear properties (such as steel yield), and the total force is plotted against a reference displacement to define a capacity curve. This can then be combined with a demand curve (typically in the form of an acceleration-displacement response spectrum (ADRS)). This essentially reduces the problem to a single degree of freedom (SDOF) system.

Nonlinear static procedures use equivalent SDOF structural models and represent seismic ground motion with response spectra. Story drifts and component actions are related subsequently to the global demand parameter by the pushover or capacity curves that are the basis of the non-linear static procedures.

Nonlinear Dynamic Analysis

Nonlinear dynamic analysis utilizes the combination of ground motion records with a detailed structural model, therefore is capable of producing results with relatively low uncertainty. In nonlinear dynamic analyses, the detailed structural model subjected to a ground-motion record produces estimates of component deformations for each degree of freedom in the model and the modal responses are combined using schemes such as the square-root-sum-of-squares.

In non-linear dynamic analysis, the non-linear properties of the structure are considered as part of a time domain analysis. This approach is the most rigorous, and is required by some building codes for buildings of unusual configuration or of special importance. However, the calculated response can be very sensitive to the characteristics of the individual ground motion used as seismic input; therefore, several analyses are required using different ground motion records to achieve a reliable estimation of the probabilistic distribution of structural response. Since the properties of the seismic response depend on the intensity, or severity, of the seismic shaking, a comprehensive assessment calls for numerous nonlinear dynamic analyses at various levels of intensity to represent different possible earthquake scenarios. This has led to the emergence of methods like the Incremental Dynamic Analysis.

Seismic Retrofit

Seismic retrofitting is the modification of existing structures to make them more resistant to seismic activity, ground motion, or soil failure due to earthquakes. With better understanding of seismic demand on structures and with our recent experiences with large earthquakes near urban centers, the need of seismic retrofitting is well acknowledged. Prior to the introduction of modern seismic codes in the late 1960s for developed countries (US, Japan etc.) and late 1970s for many other parts of the world (Turkey, China etc.), many structures were designed without adequate detailing and reinforcement for seismic protection. In view of the imminent problem, various research work has been carried out. State-of-the-art technical guidelines for seismic assessment, retrofit and rehabilitation have been published around the world - such as the ASCE-SEI 41 and the New Zealand Society for Earthquake Engineering (NZSEE)'s guidelines. These codes must be regularly updated; the 1994 Northridge earthquake brought to light the brittleness of welded steel frames, for example.

Infill shear trusses — University of California dormitory, Berkeley

The retrofit techniques outlined here are also applicable for other natural hazards such as tropical cyclones, tornadoes, and severe winds from thunderstorms. Whilst current practice of seismic retrofitting is predominantly concerned with structural improvements to reduce the seismic hazard of using the structures, it is similarly essential to reduce the hazards and losses from non-structural elements. It is also important to keep in mind that there is no such thing as an earthquake-proof structure, although seismic performance can be greatly enhanced through proper initial design or subsequent modifications.

External bracing of an existing reinforced concrete parking garage (Berkeley)

Port Authority Bus Terminal

Strategies

Seismic retrofit (or rehabilitation) strategies have been developed in the past few decades following the introduction of new seismic provisions and the availability of advanced materials (e.g. fiber-reinforced polymers (FRP), fiber reinforced concrete and high strength steel). Retrofit strategies are different from retrofit techniques, where the former is the basic approach to achieve an overall retrofit performance objective, such as increasing strength, increasing deformability, reducing deformation demands while the latter is the technical methods to achieve that strategy, for example FRP jacketing.

- Increasing the global capacity (strengthening). This is typically done by the addition of cross braces or new structural walls.

- Reduction of the seismic demand by means of supplementary damping and/or use of base isolation systems.

- Increasing the local capacity of structural elements. This strategy recognises the inherent capacity within the existing structures, and therefore adopt a more cost-effective approach to selectively upgrade local capacity (deformation/ductility, strength or stiffness) of individual structural components.

- Selective weakening retrofit. This is a counter intuitive strategy to change the inelastic mechanism of the structure, while recognising the inherent capacity of the structure.

- Allowing sliding connections such as passageway bridges to accommodate additional movement between seismically independent structures.

- Addition of seismic friction dampers to brace weak structures and provide damping.

Performance Objectives

In the past, seismic retrofit was primarily applied to achieve public safety, with engineering solutions limited by economic and political considerations. However, with the development of Performance based earthquake engineering (PBEE), several levels of performance objectives are gradually recognised:

- Public safety only. The goal is to protect human life, ensuring that the structure will not collapse upon its occupants or passersby, and that the structure can be safely exited. Under severe seismic conditions the structure may be a total economic write-off, requiring tear-down and replacement.

- Structure survivability. The goal is that the structure, while remaining safe for exit, may require extensive repair (but not replacement) before it is generally useful or considered safe for occupation. This is typically the lowest level of retrofit applied to bridges.

- Structure functionality. Primary structure undamaged and the structure is undiminished in utility for its primary application. A high level of retrofit, this ensures that any required repairs are only "cosmetic" - for example, minor cracks in plaster, drywall and stucco. This is the minimum acceptable level of retrofit for hospitals.

- Structure unaffected. This level of retrofit is preferred for historic structures of high cultural significance.

Techniques

Common seismic retrofitting techniques fall into several categories:

One of many "earthquake bolts" found throughout period houses in the city of Charleston subsequent to the Charleston earthquake of 1886. They could be tightened and loosened to support the house without having to otherwise demolish the house due to instability. The bolts were directly loosely connected to the supporting frame of the house.

External Post-tensioning

The use of external post-tensioning for new structural systems have been developed in the past decade. Under the PRESS (Precast Seismic Structural Systems), a large-scale U.S./Japan joint research program, unbonded post-tensioning high strength steel tendons have been used to achieve a moment-resisting system that has self-centering capacity. An extension of the same idea for seismic retrofitting has been experimentally tested for seismic retrofit of California bridges under a Caltrans research project and for seismic retrofit of non-ductile reinforced concrete frames. Pre-stressing can increase the capacity of structural elements such as beam, column and beam-column joints. It should be noted that external pre-stressing has been used for structural upgrade for gravity/live loading since the 1970s.

Base Isolators

Base isolation is a collection of structural elements of a building that should substantially decouple the building's structure from the shaking ground thus protecting the building's integrity and enhancing its seismic performance. This earthquake engineering technology, which is a kind of seismic vibration control, can be applied both to a newly designed building and to seismic upgrading of existing structures. Normally, excavations are made around the building and the building is separated from the foundations. Steel or reinforced concrete beams replace the connections to the foundations, while under these, the isolating pads, or base isolators, replace the material removed. While the base isolation tends to restrict transmission of the ground motion to the building, it also keeps the building positioned properly over the foundation. Careful attention to detail is required where the building interfaces with the ground, especially at entrances, stairways and ramps, to ensure sufficient relative motion of those structural elements.

Supplementary Dampers

Supplementary dampers absorb the energy of motion and convert it to heat, thus "damping" resonant effects in structures that are rigidly attached to the ground. In addition to adding energy dissipation capacity to the structure, supplementary damping can reduce the displacement and acceleration demand within the structures. In some cases, the threat of damage does not come from the initial shock itself, but rather from the periodic resonant motion of the structure that repeated ground motion induces. In the practical sense, supplementary dampers act similarly to Shock absorbers used in automotive suspensions.

Tuned Mass Dampers

Tuned mass dampers (TMD) employ movable weights on some sort of springs. These are typically employed to reduce wind sway in very tall, light buildings. Similar designs may be employed to impart earthquake resistance in eight to ten story buildings that are prone to destructive earthquake induced resonances.

Slosh Tank

A slosh tank is a large tank of fluid placed on an upper floor. During a seismic event, the fluid in this tank will slosh back and forth, but is directed by baffles - partitions that prevent the tank itself becoming resonant; through its mass the water may change or counter the resonant period of the building. Additional kinetic energy can be converted to heat by the baffles and is dissipated through the water - any temperature rise will be insignificant. One Rincon Hill in San Francisco is a recent skyscraper with a rooftop slosh tank.

Active Control System

Very tall buildings ("skyscrapers"), when built using modern lightweight materials, might sway uncomfortably (but not dangerously) in certain wind conditions. A solution to this problem is to include at some upper story a large mass, constrained, but free to move within a limited range, and moving on some sort of bearing system such as an air cushion or hydraulic film. Hydraulic pistons, powered by electric pumps and accumulators, are actively driven to counter the wind forces and natural resonances. These may also, if properly designed, be effective in controlling excessive motion - with or without applied power - in an earthquake. In general, though, modern steel frame high rise buildings are not as subject to dangerous motion as are medium rise (eight to ten story) buildings, as the resonant period of a tall and massive building is longer than the approximately one second shocks applied by an earthquake.

Adhoc Addition of Structural Support/Reinforcement

The most common form of seismic retrofit to lower buildings is adding strength to the existing structure to resist seismic forces. The strengthening may be limited to connections between existing building elements or it may involve adding primary resisting elements such as walls or frames, particularly in the lower stories.

Connections between Buildings and their Expansion Additions

Frequently, building additions will not be strongly connected to the existing structure, but simply placed adjacent to it, with only minor continuity in flooring, siding, and roofing. As a result, the addition may have a different resonant period than the original structure, and they may easily detach from one another. The relative motion will then cause the two parts to collide, causing severe structural damage. Seismic modification will either tie the two building components rigidly together so that they behave as a single mass or it will employ dampers to expend the energy from relative motion, with appropriate allowance for this motion, such as increased spacing and sliding bridges between sections.

Exterior Reinforcement of Building

Exterior Concrete Columns

Historic buildings, made of unreinforced masonry, may have culturally important interior detailing or murals that should not be disturbed. In this case, the solution may be to add a number of steel, reinforced concrete, or poststressed concrete columns to the exterior. Careful attention must be paid to the connections with other members such as footings, top plates, and roof trusses.

Infill Shear Trusses

Shown here is an exterior shear reinforcement of a conventional reinforced concrete dormitory building. In this case, there was sufficient vertical strength in the building columns and sufficient shear strength in the lower stories that only limited shear reinforcement was required to make it earthquake resistant for this location near the Hayward fault.

Massive Exterior Structure

In other circumstances, far greater reinforcement is required. In the structure shown at right — a parking garage over shops — the placement, detailing, and painting of the reinforcement becomes itself an architectural embellishment.

Typical Retrofit Scenario & Solution

Soft-story Failure

Partial failure due to inadequate shear structure at garage level. Damage in San Francisco due to the Loma Prieta event.

This collapse mode is known as *soft story collapse*. In many buildings the ground level is designed for different uses than the upper levels. Low rise residential structures may be built over a parking garage which have large doors on one side. Hotels may have a tall ground floors to allow for a grand entrance or ballrooms. Office buildings may have stores in the ground floor which desire continuous windows for display.

Traditional seismic design assumes that the lower stories of a building are stronger than the upper stories and where this is not the case—if the lower story is less strong than the upper structure—the structure will not respond to earthquakes in the expected fashion. Using modern design methods, it is possible to take a weak story into account. Several failures of this type in one large apartment complex caused most of the fatalities in the 1994 Northridge earthquake.

Typically, where this type of problem is found, the weak story is reinforced to make it stronger than the floors above by adding shear walls or moment frames. Moment frames consisting of inverted **U** bents are useful in preserving lower story garage access, while a lower cost solution may be to use shear walls or trusses in several locations, which partially reduce the usefulness for automobile parking but still allow the space to be used for other storage.

Beam-column Joint Connections

Beam-column joint connections are a common structural weakness in dealing with seismic retrofitting. Prior to the introduction of modern seismic codes in early 1970s, beam-column joints were typically non-engineered or designed. Laboratory testings have confirmed the seismic vulnerability of these poorly detailed and under-designed connections. Failure of beam-column joint connections can typically lead to catastrophic collapse of a frame-building, as often observed in recent earthquakes

For reinforced concrete beam-column joints - various retrofit solutions have been proposed and tested in the past 20 years. Philosophically, the various seismic retrofit strategies discussed above can be implemented for reinforced concrete joints. Concrete or steel jacketing have been a popular retrofit technique until the advent of composite materials such as Carbon fiber-reinforced polymer (FRP). Composite materials such as carbon FRP and aramic FRP have been extensively tested for use in seismic retrofit with some success. One novel technique includes the use of selective weakening of the beam and added external post-tensioning to the joint in order to achieve flexural hinging in the beam, which is more desirable in terms of seismic design.

Widespread weld failures at beam-column joints of low-to-medium rise steel buildings during the Northridge 1994 earthquake for example, have shown the structural defiencies of these 'modern-designed' post-1970s welded moment-resisting connections. A subsequent SAC research project has documented, tested and proposed several retrofit solutions for these welded steel moment-resisting connections. Various retrofit solutions have been developed for these welded joints - such as a) weld strengthening and b) addition of steel haunch or 'dog-bone' shape flange.

Following the Northridge earthquake, a number of steel moment -frame buildings were found to have experienced brittle fractures of beam to column connections. Discovery of these unanticipated brittle fractures of framing connections was alarming to engineers and the building industry. Starting in the 1960s, engineers began to regard welded steel moment-frame buildings as being among the most ductile systems contained in the building code. Many engineers believed that steel moment-frame buildings were essentially invulnerable to earthquake induced damage and thought that should damage occur, it would be limited to ductile yielding of members and connections. Observation of damage sustained by buildings in the 1994 Northridge earthquake indicated that contrary to the intended behavior, in many cases, brittle fractures initiated within the connections at very low levels of plastic demand. In September, 1994, The SAC joint Venture, AISC, AISI, and NIST jointly convened an international workshop in Los Angeles to coordinate the efforts of various participants and to lay the foundation for systematic investigation and resolution of the problem. In September 1995 the SAC Joint Venture entered into a contractual agreement with FEMA to conduct Phase II of the SAC Steel project. Under Phase II, SAC continued its extensive problem-focused study of the performance of moment resisting steel frames and connections of various configurations, with the ultimate goal of developing seismic design criteria for steel construction. As a result of these studies it is now known that the typical moment-resisting connection detail employed in steel moment frame construction prior to the 1994 Northridge earthquake had a number of features that rendered it inherently susceptible to brittle fracture.

Shear Failure within Floor Diaphragm

Floors in wooden buildings are usually constructed upon relatively deep spans of wood, called joists, covered with a diagonal wood planking or plywood to form a subfloor upon which the finish floor surface is laid. In many structures these are all aligned in the same direction. To prevent the beams from tipping over onto their side, blocking is used at each end, and for additional stiffness, blocking or diagonal wood or metal bracing may be placed between beams at one or more points in their spans. At the outer edge it is typical to use a single depth of blocking and a perimeter beam overall.

If the blocking or nailing is inadequate, each beam can be laid flat by the shear forces applied to the building. In this position they lack most of their original strength and the structure may further collapse. As part of a retrofit the blocking may be doubled, especially at the outer edges of the building. It may be appropriate to add additional nails between the sill plate of the perimeter wall erected upon the floor diaphragm, although this will require exposing the sill plate by removing interior plaster or exterior siding. As the sill plate may be quite old and dry and substantial nails must be used, it may be necessary to pre-drill a hole for the nail in the old wood to avoid splitting. When the wall is opened for this purpose it may also be appropriate to tie vertical wall elements into the foundation using specialty connectors and bolts glued with epoxy cement into holes drilled in the foundation.

Sliding off Foundation and "Cripple Wall" Failure

Single or two story wood-frame domestic structures built on a perimeter or slab foundation are relatively safe in an earthquake, but in many structures built before 1950 the sill plate that sits between the concrete foundation and the floor diaphragm (perimeter foundation) or studwall (slab foundation) may not be sufficiently bolted in. Additionally, older attachments (without substantial corrosion-proofing) may have corroded to a point of weakness. A sideways shock can slide the building entirely off of the foundations or slab.

House slid off of foundation

Low cripple wall collapse and detachment of structure from concrete stairway

Often such buildings, especially if constructed on a moderate slope, are erected on a platform connected to a perimeter foundation through low stud-walls called "cripple wall" or *pin-up*. This low wall structure itself may fail in shear or in its connections to itself at the corners, leading to the building moving diagonally and collapsing the low walls. The likelihood of failure of the pin-up can be reduced by ensuring that the corners are well reinforced in shear and that the shear panels are well connected to each other through the corner posts. This requires structural grade sheet plywood, often treated for rot resistance. This grade of plywood is made without interior unfilled knots and with more, thinner layers than common plywood. New buildings designed to resist earthquakes will typically use OSB (oriented strand board), sometimes with metal joins between panels, and with well attached stucco covering to enhance its performance. In many modern tract homes, especially those built upon expansive (clay) soil the building is constructed upon a single and relatively thick monolithic slab, kept in one piece by high tensile rods that are stressed after the slab has set. This poststressing places the concrete under compression - a condition under which it is extremely strong in bending and so will not crack under adverse soil conditions.

Multiple Piers in Shallow Pits

Some older low-cost structures are elevated on tapered concrete pylons set into shallow pits, a method frequently used to attach outdoor decks to existing buildings. This is seen in conditions of damp soil, especially in tropical conditions, as it leaves a dry ventilated space under the house, and in far northern conditions of permafrost (frozen mud) as it keeps the building's warmth from destabilizing the ground beneath. During an earthquake, the pylons may tip, spilling the building to the ground. This can be overcome by using deep-bored holes to contain cast-in-place reinforced pylons, which are then secured to the floor panel at the corners of the building. Another technique is to add sufficient diagonal bracing or sections of concrete shear wall between pylons.

Reinforced Concrete Column Burst

Jacketed and grouted column on left, unmodified on right

Reinforced concrete columns typically contain large diameter vertical rebar (reinforcing bars) arranged in a ring, surrounded by lighter-gauge hoops of rebar. Upon analysis of failures due to earthquakes, it has been realized that the weakness was not in the vertical bars, but rather in inadequate strength and quantity of hoops. Once the integrity of the hoops is breached, the vertical rebar can flex outward, stressing the central column of concrete. The concrete then simply crumbles into small pieces, now unconstrained by the surrounding rebar. In new construction a greater amount of hoop-like structures are used.

One simple retrofit is to surround the column with a jacket of steel plates formed and welded into a single cylinder. The space between the jacket and the column is then filled with concrete, a process called grouting. Where soil or structure conditions require such additional modification, additional pilings may be driven near the column base and concrete pads linking the pilings to the pylon are fabricated at or below ground level. In the example shown not all columns needed to be modified to gain sufficient seismic resistance for the conditions expected. (This location is about a mile from the Hayward Fault Zone.)

Reinforced Concrete Wall Burst

Concrete walls are often used at the transition between elevated road fill and overpass structures. The wall is used both to retain the soil and so enable the use of a shorter span and also to transfer the weight of the span directly downward to footings in undisturbed soil. If these walls are inadequate they may crumble under the stress of an earthquake's induced ground motion.

One form of retrofit is to drill numerous holes into the surface of the wall, and secure short L-shaped sections of rebar to the surface of each hole with epoxy adhesive. Additional vertical and horizontal rebar is then secured to the new elements, a form is erected, and an additional layer of concrete is poured. This modification may be combined with additional footings in excavated trenches and additional support ledgers and tiebacks to retain the span on the bounding walls.

Brick Wall Resin and Glass Fiber Reinforcement

Brick building structures have been reinforced with coatings of glass fiber and appropriate resin (epoxy or polyester). In lower floors these may be applied over entire exposed surfaces, while in upper floors this may be confined to narrow areas around window and door openings. This application provides tensile strength that stiffens the wall against bending away from the side with the application. The efficient protection of an entire building requires extensive analysis and engineering to determine the appropriate locations to be treated.

Lift

Where moist or poorly consolidated alluvial soil interfaces in a "beach like" structure against underlying firm material, seismic waves traveling through the alluvium can be amplified, just as are water waves against a sloping beach. In these special conditions, vertical accelerations up to twice the force of gravity have been measured. If a building is not secured to a well-embedded foundation it is possible for the building to be thrust from (or with) its foundations into the air, usually with severe damage upon landing. Even if it is well-founded, higher portions such as upper stories or roof structures or attached structures such as canopies and porches may become detached from the primary structure.

Good practices in modern, earthquake-resistant structures dictate that there be good vertical connections throughout every component of the building, from undisturbed or engineered earth to foundation to sill plate to vertical studs to plate cap through each floor and continuing to the roof structure. Above the foundation and sill plate the connections are typically made using steel strap or sheet stampings, nailed to wood members using special hardened high-shear strength nails, and heavy angle stampings secured with through bolts, using large washers to prevent pull-through. Where inadequate bolts are provided between the sill plates and a foundation in existing construction (or are not trusted due to possible corrosion), special clamp plates may be added, each of which is secured to the foundation using expansion bolts inserted into holes drilled in an exposed face of concrete. Other members must then be secured to the sill plates with additional fittings.

Soil

One of the most difficult retrofits is that required to prevent damage due to soil failure. Soil failure can occur on a slope, a slope failure or landslide, or in a flat area due to liquefaction of water-saturated sand and/or mud. Generally, deep pilings must be driven into stable soil (typically hard mud or sand) or to underlying bedrock or the slope must be stabilized. For buildings built atop previous landslides the practicality of retrofit may be limited by economic factors, as it is not practical to stabilize a large, deep landslide. The likelihood of landslide or soil failure may also depend upon seasonal factors, as the soil may be more stable at the beginning of a wet season than at the beginning of the dry season. Such a "two season" *Mediterranean climate* is seen throughout California.

In some cases, the best that can be done is to reduce the entrance of water runoff from higher, stable elevations by capturing and bypassing through channels or pipes, and to drain water infiltrated directly and from subsurface springs by inserting horizontal perforated tubes. There are numerous locations in California where extensive developments have been built atop archaic landslides, which have not moved in historic times but which (if both water-saturated and shaken by an earthquake) have a high probability of moving *en masse*, carrying entire sections of suburban development to new locations. While the most modern of house structures (well tied to monolithic concrete foundation slabs reinforced with post tensioning cables) may survive such movement largely intact, the building will no longer be in its proper location.

Utility Pipes and Cables: Risks

Natural gas and propane supply pipes to structures often prove especially dangerous during and after earthquakes. Should a building move from its foundation or fall due to cripple wall collapse, the ductile iron pipes transporting the gas within the structure may be broken, typically at the location of threaded joints. The gas may then still be provided to the pressure regulator from higher pressure lines and so continue to flow in substantial quantities; it may then be ignited by a nearby source such as a lit pilot light or arcing electrical connection.

There are two primary methods of automatically restraining the flow of gas after an earthquake, installed on the low pressure side of the regulator, and usually downstream of the gas meter.

- A caged metal ball may be arranged at the edge of an orifice. Upon seismic shock, the ball will roll into the orifice, sealing it to prevent gas flow. The ball may later be reset by the use of an external magnet. This device will respond only to ground motion.

- A flow-sensitive device may be used to close a valve if the flow of gas exceeds a set threshold (very much like an electrical circuit breaker). This device will operate independently of seismic motion, but will not respond to minor leaks which may be caused by an earthquake.

It appears that the most secure configuration would be to use one of each of these devices in series.

Tunnels

Unless the tunnel penetrates a fault likely to slip, the greatest danger to tunnels is a landslide blocking an entrance. Additional protection around the entrance may be applied to divert any falling material (similar as is done to divert snow avalanches) or the slope above the tunnel may be stabilized in some way. Where only small- to medium-sized rocks and boulders are expected to fall, the entire slope may be covered with wire mesh, pinned down to the slope with metal rods. This is also a common modification to highway cuts where appropriate conditions exist.

Underwater Tubes

The safety of underwater tubes is highly dependent upon the soil conditions through which the tunnel was constructed, the materials and reinforcements used, and the maximum predicted earthquake expected, and other factors, some of which may remain unknown under current knowledge.

BART Tube

A tube of particular structural, seismic, economic, and political interest is the BART (Bay Area Rapid Transit) transbay tube. This tube was constructed at the bottom of San Francisco Bay through an innovative process. Rather than pushing a shield through the soft bay mud, the tube was constructed on land in sections. Each section consisted of two inner train tunnels of circular cross section, a central access tunnel of rectangular cross section, and an outer oval shell encompassing the three inner tubes. The intervening space was filled with concrete. At the bottom of the bay a trench was excavated and a flat bed of crushed stone prepared to receive the tube sections. The sections were then floated into place and sunk, then joined with bolted connections to previously-placed sections. An overfill was then placed atop the tube to hold it down. Once completed from San Francisco to Oakland, the tracks and electrical components were installed. The predicted response of the tube during a major earthquake was likened to be as that of a string of (cooked) spaghetti in a bowl of gelatin dessert. To avoid overstressing the tube due to differential movements at each end, a sliding slip joint was included at the San Francisco terminus under the landmark Ferry Building.

The engineers of the construction consortium PBTB (Parsons Brinckerhoff-Tudor-Bechtel) used the best estimates of ground motion available at the time, now known to be insufficient given modern computational analysis methods and geotechnical knowledge. Unexpected settlement of the tube has reduced the amount of slip that can be accommodated without failure. These factors have resulted in the slip joint being designed too short to ensure survival of the tube under possible (perhaps even likely) large earthquakes in the region. To correct this deficiency the slip joint must be extended to allow for additional movement, a modification expected to be both expensive and technically and logistically difficult. Other retrofits to the BART tube include vibratory consolidation of the tube's overfill to avoid potential liquefying of the overfill, which has now been completed. (Should the overfill fail there is a danger of portions of the tube rising from the bottom, an event which could potentially cause failure of the section connections.)

Bridge Retrofit

Bridges have several failure modes.

Expansion Rockers

Many short bridge spans are statically anchored at one end and attached to rockers at the other. This rocker gives vertical and transverse support while allowing the bridge span to expand and contract with temperature changes. The change in the length of the span is accommodated over a gap in the roadway by comb-like expansion joints. During severe ground motion, the rockers may jump from their tracks or be moved beyond their design limits, causing the bridge to unship from its resting point and then either become misaligned or fail completely. Motion can be constrained by adding ductile or high-strength steel restraints that are friction-clamped to beams and designed to slide under extreme stress while still limiting the motion relative to the anchorage.

Deck Rigidity

Suspension bridges may respond to earthquakes with a side-to-side motion exceeding that which was designed for wind gust response. Such motion can cause fragmentation of the road surface, damage to bearings, and plastic deformation or breakage of components. Devices such as hydraulic dampers or clamped sliding connections and additional diagonal reenforcement may be added.

Lattice Girders, Beams, and Ties

Obsolete riveted lattice members

Lattice girders consist of two "I"-beams connected with a criss-cross lattice of flat strap or angle stock. These can be greatly strengthened by replacing the open lattice with plate members. This is usually done in concert with the replacement of hot rivets with bolts.

Bolted plate lattice replacement, forming box members

Hot Rivets

Many older structures were fabricated by inserting red-hot rivets into pre-drilled holes; the soft rivets are then peened using an air hammer on one side and a bucking bar on the head end. As these cool slowly, they are left in an annealed (soft) condition, while the plate, having been hot rolled and quenched during manufacture, remains relatively hard. Under extreme stress the hard plates can shear the soft rivets, resulting in failure of the joint.

The solution is to burn out each rivet with an oxygen torch. The hole is then prepared to a precise diameter with a reamer. A special *locator bolt*, consisting of a head, a shaft matching the reamed hole, and a threaded end is inserted and retained with a nut, then tightened with a wrench. As the bolt has been formed from an appropriate high-strength alloy and has also been heat-treated, it is not subject to either the plastic shear failure typical of hot rivets nor the brittle fracture of ordinary bolts. Any partial failure will be in the plastic flow of the metal secured by the bolt; with proper engineering any such failure should be non-catastrophic.

Fill and Overpass

Elevated roadways are typically built on sections of elevated earth fill connected with bridge-like segments, often supported with vertical columns. If the soil fails where a bridge terminates, the bridge may become disconnected from the rest of the roadway and break away. The retrofit for this is to add additional reinforcement to any support-ing wall, or to add deep caissons adjacent to the edge at each end and connect them with a supporting beam under the bridge.

Another failure occurs when the fill at each end moves (through resonant effects) in bulk, in opposite directions. If there is an insufficient founding shelf for the overpass, then it may fall. Additional shelf and ductile stays may be added to attach the overpass to the footings at one or both ends. The stays, rather than being fixed to the beams, may instead be clamped to them. Under moderate loading, these keep the overpass centered in the gap so that it is less likely to slide off its founding shelf at one end. The ability for the fixed ends to slide, rather than break, will prevent the complete drop of the struc-ture if it should fail to remain on the footings.

Viaducts

Cypress Freeway viaduct collapse. Note failure of inadequate anti-burst wrapping and lack of connection between upper and lower vertical elements.

Large sections of roadway may consist entirely of viaduct, sections with no connection to the earth other than through vertical columns. When concrete columns are used, the detailing is critical. Typical failure may be in the toppling of a row of columns due either to soil connection failure or to insufficient cylindrical wrapping with rebar. Both failures were seen in the 1995 Great Hanshin earthquake in Kobe, Japan, where an entire viaduct, centrally supported by a single row of large columns, was laid down to one side. Such columns are reinforced by excavating to the foundation pad, driving additional pilings, and adding a new, larger pad, well connected with rebar alongside or into the column. A column with insufficient wrapping bar, which is prone to burst and then hinge at the bursting point, may be completely encased in a circular or elliptical jacket of welded steel sheet and grouted as described above.

Sometimes viaducts may fail in the connections between components. This was seen in the failure of the Cypress Freeway in Oakland, California, during the Loma Prieta earthquake. This viaduct was a two-level structure, and the upper portions of the columns were not well connected to the lower portions that supported the lower level; this caused the upper deck to collapse upon the lower deck. Weak connections such as these require additional external jacketing - either through external steel components or by a complete jacket of reinforced concrete, often using stub connections that are glued (using epoxy adhesive) into numerous drilled holes. These stubs are then connected to additional wrappings, external forms (which may be temporary or permanent) are erected, and additional concrete is poured into the space. Large connected structures similar to the Cypress Viaduct must also be properly analyzed in their entirety using dynamic computer simulations.

Residential Retrofit

Side-to-side forces cause most earthquake damage. Bolting of the mudsill to the foundation and application of plywood to cripple walls are a few basic retrofit techniques which homeowners may apply to wood-framed residential structures to mitigate the effects of seismic activity. The City of San Leandro created guidelines for these procedures, as outlined in the following pamphlet. Public awareness and initiative are critical to the retrofit and preservation of existing building stock, and such efforts as those of the Association of Bay Area Governments are instrumental in providing informational resources to seismically active communities.

Wood Frame Structure

Most houses in North America are wood-framed structures. Wood is one of the best materials for earthquake-resistant construction since it is lightweight and more flexible than masonry. It is easy to work with and less expensive than steel, masonry, or concrete. In older homes the most significant weaknesses are the connection from the wood-framed walls to the foundation and the relatively weak "cripple-walls." (Cripple walls are the short wood walls that extend from the top of the foundation to the lowest floor level in houses that have raised floors.) Adding connections from the base of the wood-framed structure to the foundation is almost always an important part of a seismic retrofit. Bracing the cripple-walls to resist side-to-side forces is essential in houses with cripple walls; bracing is usually done with plywood. Oriented strand board (OSB) does not perform as consistently as plywood, and is not the favored choice of retrofit designers or installers.

Retrofit methods in older woodframe structures may consist of the following, and other methods not described here.

- The lowest plate rails of walls (usually called "mudsills" or "foundation sills" in North America) are bolted to a continuous foundation, or secured with rigid metal connectors bolted to the foundation so as to resist side-to-side forces.

- *Cripple walls* are braced with plywood.

- Selected vertical elements (typically the posts at the ends of plywood wall bracing panels) are connected to the foundation. These connections are intended to prevent the braced walls from rocking up and down when subjected to back-and-forth forces at the top of the braced walls, not to resist the wall or house "jumping" off the foundation (which almost never occurs).

- In two story buildings using "platform framing" (sometimes called "western" style construction, where walls are progressively erected upon the lower story's upper diaphragm, unlike "eastern" or *balloon framing*), the upper walls are connected to the lower walls with tension elements. In some cases, connections may be extended vertically to include retention of certain roof elements. This sort of strengthening is usually very costly with respect to the strength gained.

- Vertical posts are secured to the beams or other members they support. This is particularly important where loss of support would lead to collapse of a segment of a building. Connections from posts to beams cannot resist appreciable side-to-side forces; it is much more important to strengthen around the perimeter of a building (bracing the cripple-walls and supplementing foundation-to-wood-framing connections) than it is to reinforce post-to-beam connections.

Wooden framing is efficient when combined with masonry, if the structure is properly designed. In Turkey, the traditional houses (bagdadi) are made with this technology. In El Salvador, wood and bamboo are used for residential construction.

Reinforced and Unreinforced Masonry

In many parts of developing countries such as Pakistan, Iran and China, unreinforced or in some cases reinforced masonry is the predominantly form of structures for rural residential and dwelling. Masonry was also a common construction form in the early part of the 20th century, which implies that a substantial number of these at-risk masonry structures would have significant heritage value. Masonry walls that are not reinforced are especially hazardous. Such structures may be more appropriate for replacement than retrofit, but if the walls are the principal load bearing elements in structures of modest size they may be appropriately reinforced. It is especially important that floor and ceiling beams be securely attached to the walls. Additional vertical supports in the form of steel or reinforced concrete may be added.

In the western United States, much of what is seen as masonry is actually brick or stone veneer. Current construction rules dictate the amount of tie–back required, which consist of metal straps secured to vertical structural elements. These straps extend into mortar courses, securing the veneer to the primary structure. Older structures may not secure this sufficiently for seismic safety. A weakly secured veneer in a house interior (sometimes used to face a fireplace from floor to ceiling) can be especially dangerous to occupants. Older masonry chimneys are also dangerous if they have substantial vertical extension above the roof. These are prone to breakage at the roofline and may fall into the house in a single large piece. For retrofit, additional supports may be added; however, it is extremely expensive to strengthen an existing masonry chimney to conform with contemporary design standards. It is best to simply remove the extension and replace it with lighter materials, with special metal flue replacing the flue tile and a wood structure replacing the masonry. This may be matched against existing brickwork by using very thin veneer (similar to a tile, but with the appearance of a brick).

Mitigation of Seismic Motion

Mitigation of seismic motion is an important factor in earthquake engineering and construction in earthquake-prone areas. The destabilizing action of an earthquake on constructions may be direct (seismic motion of the ground) or indirect (earthquake-induced landslides, liquefaction of the foundation soils and waves of tsunami).

Knowledge of local amplification of the seismic motion from the bedrock is very important in order to choose the suitable design solutions. Local amplification can be anticipated from the presence of particular stratigraphic conditions, such as soft soil overlapping the bedrock, or where morphological settings (e.g. crest zones, steep slopes, valleys, or endorheic basins) may produce focalization of the seismic event.

The identification of the areas potentially affected by earthquake-induced landslides and by soil liquefaction can be made by geological survey and by analysis of historical documents. Even quiescent and stabilized landslide areas may be reactivated by severe earthquake. Young soil may be particularly susceptible to liquefaction.

Earthquake Network

Earthquake Network is a research project which aims at developing and maintaining a crowdsourced smartphone-based earthquake warning system at a global level. Smartphones made available by the population are used to detect the earthquake waves using the on-board accelerometers. When an earthquake is detected, an earthquake warning is issued in order to alert the population not yet reached by the damaging waves of the earthquake.

The project started on January 1, 2013 with the release of the homonymous Android application Earthquake Network. The author of the research project and developer of the smartphone application is Francesco Finazzi of the University of Bergamo, Italy.

Scientific Research

Earthquake warning systems are intended to rapidly detect earthquakes in order to alert the population in advance. When an earthquake is detected, a potentially large amount of people who will be affected in locations not too close to the epicenter can receive the warning several seconds (10 to 60) before damaging shaking occurs. This is possible since the warning can be delivered at a higher speed than the speed of the earthquake waves. The Earthquake Network project focuses on developing an earthquake warning system using smartphones rather than professional seismometers.

Working Principle

Smartphones with the Earthquake Network application installed are nodes of the sensor network of the Earthquake Network project. When a smartphone is not in use and it is connected to a source of power, the application switches on the accelerometer in order to read the smartphone acceleration. If a threshold is exceeded, the smartphone sends a signal to a central server. The server collects the signals sent by all the smartphones and, thanks to a statistical algorithm, it decides in real time whether an earthquake is likely occurring. If an earthquake is detected, the server instantly notifies all the smartphones with the application installed. An alarm goes off when the notification is received, and the smartphone owner can take cover.

Network Size and Geographic Distribution

The number of smartphones in the network is highly variable as users can install or uninstall the Earthquake Network application at any time. Additionally, the number of active smartphones (not in use and connected to a source of power) constantly changes during the day. Globally, the total number of smartphones with the application installed is around 129,000 (April 2016) and the number of active smartphones ranges from around 2,000 to around 8,000 depending on the hour of the day. The geographic distribution of the network nodes is given in the following table.

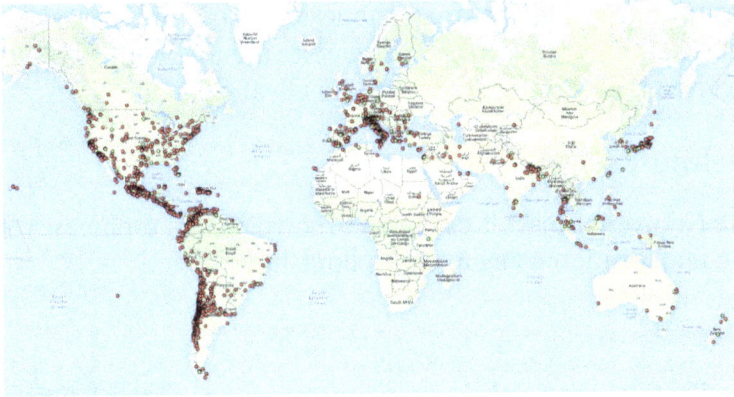

distribution Smartphone network spatial (green and red dots) on December 4, 2015

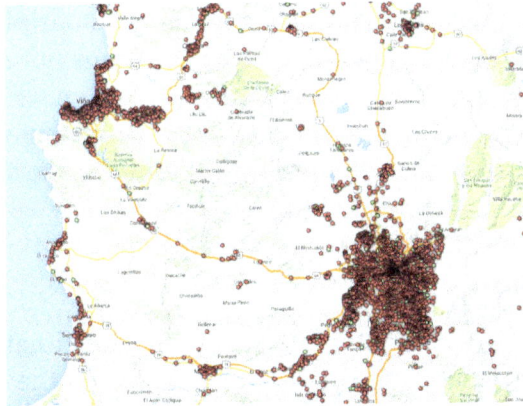

Smartphone network in Santiago area on December 5, 2015

Geographic distribution (April 2016)	
Country	**Nodes (%)**
Ecuador	41.4
Chile	31.9
Mexico	7.8
Argentina	2.7
United States	2.5
Italy	2.4
Colombia	2.3
Peru	1.3
Others	7.7

Detected Earthquakes

From the beginning of the project, the smartphone network detected 239 earthquakes (April 26, 2016). Most of the earthquakes were detected in Chile where the network is quite stable in terms of number of smartphones. As an example, the magnitude 8.3 Illapel earthquake was detected by the smartphones in the city of Valparaíso. Smartphones in Santiago received the warning 10 seconds before the earthquake, while smartphones in Mendoza received the warning 20 seconds in advance.

Project Development

The Earthquake Network project is expected to solve 4 main problems related to earthquake detection and location using a smartphone network.

Real-time Detection

In order to be effective, the earthquake warning system of the Earthquake Network project is expected to detect the earthquake as fast as possible. Earthquake detection is performed by the real-time analysis of the data that the smartphone network sends to the central server. Since smartphones detect accelerations not necessarily induced by an earthquake, the server implements a statistical algorithm which is able to recognize real earthquakes from the background noise. The statistical methodology at the basis of the algorithm allows to control the probability of false alarm. Development stage: released.

Epicenter Estimation

When detection occurs, it is important to obtain an estimate of the epicenter in order to locate the geographic areas that was affected by the earthquake. Two epicenter estimation algorithms for crowdsourced smartphone-based earthquake early warning systems have been developed and they are detailed in a paper published on the Bulletin of the Seismological Society of America journal. Development stage: released.

Peak Ground Acceleration

The accelerations recorded by a dense smartphone network may be used to produce high resolution peak ground acceleration maps for the detected earthquakes. The task is complicated by the fact that smartphones are not secured to the ground and they are not directly measuring the ground acceleration. If properly calibrated, however, data from a large number of smartphones may allow to estimate the ground acceleration. Development stage: analysis.

Magnitude Estimation

The earthquake magnitude is an important parameter as it defines the energy released by the earthquake event and it allows to evaluate the severity of the earthquake in terms of potential damages to property and people. Magnitude estimation using the data collected by the smartphone network is currently under study. Development stage: analysis.

ShakeAlert

Map showing the amount of advance warning time that might be available from ShakeAlert for several plausible future earthquake scenarios.

ShakeAlert is an experimental earthquake early warning system (EEW) for the West Coast of the United States and the Pacific Northwest sponsored by the United States Geological Survey (USGS).

The system uses sensors to detect P waves, compressional primary waves created by earthquakes and traveling faster than the damaging S waves. ShakeAlert is currently in a demonstration phase and has been sending real-time alerts to selected beta users since January 2012. The system will issue automated alerts to give people time to take protective actions like "drop, cover and hold on" in the event of a quake, preventing injuries caused by falling debris, automatically stopping public transport systems, preventing cars from entering bridges or tunnels, automatically shutting down industrial systems and gas lines, and triggering specific protocols in hospitals and other sensitive work environments. Initially the system will cover the west coast of in North America which is exposed to significant seismic risk along the San Andreas fault zone or the Cascadia subduction zone.

The second algorithm crucial to ShakeAlert is the tc-Pd Onsite algorithm. By using amplitude pd and period tc of the first signs of shaking, this onsite algorithm more accurately predicts the intensity and size of the earthquake than the ElarmS do. The tradeoff for using these algorithms for the earliest detection possible means having a less reliable approach than regional warning algorithms, however some argue that the added seconds to prepare are more important than reliability. Lastly, the Virtual Seismologist, as known as the VS method, imitates the analysis of a human scientist in terms of capacity, but does so at a faster rate. A Bayesian framework is used with inputs of acceleration, velocity, and displacement. The last step requires of all these algorithms to come together in a decision module. This decision module broadcasts the probability, size, and other characteristics of the earthquake.

The ShakeAlert system alerts the surrounding populations of an earthquake that has already happened. When the system detects the earthquake, the first step is to release several horn noises at a high volume level. Next, an automated voice yells out the words earthquake. The shaking that follows can start after varying intervals of time- anywhere from a couple seconds to several minutes. The purpose of then ShakeAlert system is not to predict an earthquake before it happens, but instead to determine and locate the earthquake as soon as it begins to happen. Within a single second, ShakeAlert can detect the location and severity of the earthquake to warn people of its presence. The people that these type of signals reach can vary. In the state of California, where earthquakes happen quite often, some of those most important listeners of this system include the Bay area's rapid transit system (BART). Other important organizations that need to know the early onset of earthquakes are Fire Departments all along the west coast, as well as places like Disneyland, where a large population of people is limited to a relatively small geographical space. BART is already directly connected to the ShakeAlert system, so if the system were to send out early warnings of an earthquake, the metro system will automatically pause and adjust its trains to prevent serious accidents or derailments. Doug Given, one of the main leaders of the project, also serves as the National Earthquake Early Warning Coordinator for the U.S. Geological Survey. He explains in an interview with National Public Radio, the way in which the system is designed to alert the public. He explains that the system plans to reach populations in a similar way that the public receives Amber alerts.

Performance

The ShakeAlert system issued alerts for several significant southern California earthquakes in 2014 including a M4.4 event in Encino, a M4.2 event in Westwood, and a M5.1 event in La Habra.

ShakeAlert issued a warning 5.4 seconds after the beginning of the 6.0 magnitude earthquake that hit the Napa region on August 24, 2014. Although it was initially reported that the system provided 10 seconds of warning before the S-wave arrived in Berkeley, subsequent information showed that this was in error and the warning arrived only 5 seconds before the S-wave in Berkeley. This means the S-waves had already arrived in Napa and Vallejo when the warning was issued. San Francisco received 8 seconds warning.

Participants and Funding

The project is being developed by a consortium of institutions including the United States Geological Survey, the California Governor's Office of Emergency Services (Cal OES), the California Geological Survey, California Institute of Technology, the Berkeley Seismological Laboratory at University of California Berkeley, University of Washington, University of Oregon, and the Swiss Federal Institute of Technology in Zurich (ETHZ). The Gordon and Betty Moore Foundation has invested more than $6 million in developing the system.

USGS estimates the west coast system will cost $38 million to complete and $16 million per year to operate over and above the investment that is already made in earthquake monitoring. More than 30 Congress members have signed a joint letter urging the President to add full funding for the system to his federal budget request.

ShakeAlert warnings will be sent out via cell phones, television, radio, and even may be implemented into large buildings and malls. ShakeAlert is capable of detecting earthquakes at an early stage because of three specific algorithms. The first algorithm is ElarmS. Also known as Earthquake Alarm Systems, these signals detect the P-wave energy released from any given earthquake. This energy, while given off quite early, does not usually cause damage (Wenzel). It is also the ElarmS that are responsible for roughly estimating the geographical location and size of the earthquake. Following these Elarms, empirical attenuation relations estimate how much the earth will shake in the specified region of the quake.

In December 2014 $5 million was added to the USGS budget for ShakeAlert development. This enabled USGS to purchase $1 million in seismic instrumentation and award $4 million in funding to the project partners to make current "ShakeAlert" demonstration system more robust. As of August 2015, organizations enrolled in the beta test user program include: CalOES Warning Center, LA County Fire, LA City OEM, Amgen Corp, LADWP, Metrolink, Caltrans and Disneyland.

User Display

The User Display interprets the results of the Decision Module and rearranges the information into more user-friendly terms for the ordinary person. Some of the things that this User Display will include is: - "display of estimated magnitude" - "display of the probability of correct alarm" - "siren and audio announcement" - "calculation and display of remaining warning time for a given user"

Seismometer

Seismometers are instruments that measure motion of the ground, including those of seismic waves generated by earthquakes, volcanic eruptions, and other seismic sources. Records of seismic waves allow seismologists to map the interior of the Earth, and locate and measure the size of these different sources.

Japan Meteorological Agency Optical Electromagnetic Seismometer

Seismograph is another Greek term from *seismós* and, *gráphō*, to draw. It is often used to mean *seismometer*, though it is more applicable to the older instruments in which the measuring and recording of ground motion were combined than to modern systems, in which these functions are separated. Both types provide a continuous record of ground motion; this distinguishes them from seismoscopes, which merely indicate that motion has occurred, perhaps with some simple measure of how large it was.

The concerning technical discipline is called seismometry, a branch of seismology.

Basic Principles

A simple seismometer that is sensitive to up-down motions of the earth can be understood by visualizing a weight hanging on a spring. The spring and weight are suspended from a frame that moves along with the earth's surface. As the earth moves, the relative motion between the weight and the earth provides a measure of the vertical ground motion. If a recording system is installed, such as a rotating drum attached to the frame, and a pen attached to the mass, this relative motion between the weight and earth can be recorded to produce a history of ground motion, called a seismogram.

Any movement of the ground moves the frame. The mass tends not to move because of its inertia, and by measuring the movement between the frame and the mass, the motion of the ground can be determined.

Early seismometers used optical levers or mechanical linkages to amplify the small motions involved, recording on soot-covered paper or photographic paper. Modern instruments use electronics. In some systems, the mass is held nearly motionless relative to the frame by an electronic negative feedback loop. The motion of the mass relative to the frame is measured, and the feedback loop applies a magnetic or electrostatic force to keep the mass nearly motionless. The voltage needed to produce this force is the output of the seismometer, which is recorded digitally. In other systems the weight is allowed to move, and its motion produces a voltage in a coil attached to the mass and moving through the magnetic field of a magnet attached to the frame. This design is often used in the geophones used in seismic surveys for oil and gas.

Professional seismic observatories usually have instruments measuring three axes: north-south (y-axis), east-west (x-axis), and the vertical (z-axis). If only one axis is measured, this is usually the vertical because it is less noisy and gives better records of some seismic waves.

The foundation of a seismic station is critical. A professional station is sometimes mounted on bedrock. The best mountings may be in deep boreholes, which avoid thermal effects, ground noise and tilting from weather and tides. Other instruments are often mounted in insulated enclosures on small buried piers of unreinforced concrete. Reinforcing rods and aggregates would distort the pier as the temperature changes. A site is always surveyed for ground noise with a temporary installation before pouring the pier and laying conduit. Originally, European seismographs were placed in a particular area after a destructive earthquake. Today, they are spread to provide appropriate coverage (in the case of weak-motion seismology) or concentrated in high-risk regions (strong-motion seismology).

History

Ancient Era

In AD 132, Zhang Heng of China's Han dynasty invented the first seismoscope (by the definition above), which was called *Houfeng Didong Yi* (translated as, "instrument for measuring the seasonal winds and the movements of the Earth"). The description we have, from the History of the Later Han Dynasty, says that it was a large bronze vessel, about 2 meters in diameter; at eight points around the top were dragon's heads holding bronze balls. When there was an earthquake, one of the mouths would open and drop its ball into a bronze toad at the base, making a sound and supposedly showing the direction of the earthquake. On at least one occasion, probably at the time of a large earthquake in Gansu in AD 143, the seismoscope indicated an earthquake even though one was not felt. The available text says that inside the vessel was a central column that could move along eight tracks; this is thought to refer to a pendulum, though it is not known exactly how this was linked to a mechanism that would open only one dragon's mouth. The first ever earthquake recorded by this seismoscope was supposedly *somewhere in the east*. Days later, a rider from the east reported this earthquake.

Replica of Zhang Heng's seismoscope *Houfeng Didong Yi*

Modern Designs

The principle can be shown by an early special purpose seismometer. This consisted of a large stationary pendulum, with a stylus on the bottom. As the earth starts to move, the heavy mass of the pendulum has the inertia to stay still in the non-earth frame of reference. The result is that the stylus scratches a pattern corresponding with the Earth's movement. This type of strong motion seismometer recorded upon a smoked glass (glass with carbon soot). While not sensitive enough to detect distant earthquakes, this instrument could indicate the direction of the pressure waves and thus help find the epicenter of a local earthquake – such instruments were useful in the analysis of the 1906 San Francisco earthquake. Further re-analysis was performed in the 1980s using these early recordings, enabling a more precise determination of the initial fault break location in Marin county and its subsequent progression, mostly to the south.

Milne horizontal pendulum seismometer. One of the Important Cultural Properties of Japan. Exhibit in the National Museum of Nature and Science, Tokyo, Japan.

After 1880, most seismometers were descended from those developed by the team of John Milne, James Alfred Ewing and Thomas Gray, who worked in Japan from 1880 to 1895. These seismometers used damped horizontal pendulums. After World War II, these were adapted into the widely used Press-Ewing seismometer.

Later, professional suites of instruments for the worldwide standard seismographic network had one set of instruments tuned to oscillate at fifteen seconds, and the other at ninety seconds, each set measuring in three directions. Amateurs or observatories with limited means tuned their smaller, less sensitive instruments to ten seconds. The basic damped horizontal pendulum seismometer swings like the gate of a fence. A heavy weight is mounted on the point of a long (from 10 cm to several meters) triangle, hinged at its vertical edge. As the ground moves, the weight stays unmoving, swinging the "gate" on the hinge.

The advantage of a horizontal pendulum is that it achieves very low frequencies of oscillation in a compact instrument. The "gate" is slightly tilted, so the weight tends to slowly return to a central position. The pendulum is adjusted (before the damping is installed) to oscillate once per three seconds, or once per thirty seconds. The general-purpose instruments of small stations or amateurs usually oscillate once per ten seconds. A pan of oil is placed under the arm, and a small sheet of metal mounted on the underside of the arm drags in the oil to damp oscillations. The level of oil, position on the arm, and angle and size of sheet is adjusted until the damping is "critical," that is, almost having oscillation. The hinge is very low friction, often torsion wires, so the only friction is the internal friction of the wire. Small seismographs with low proof masses are placed in a vacuum to reduce disturbances from air currents.

Zollner described torsionally suspended horizontal pendulums as early as 1869, but developed them for gravimetry rather than seismometry.

Early seismometers had an arrangement of levers on jeweled bearings, to scratch smoked glass or paper. Later, mirrors reflected a light beam to a direct-recording plate or roll of photographic paper. Briefly, some designs returned to mechanical movements to save money. In mid-twentieth-century systems, the light was reflected to a pair of differential electronic photosensors called a photomultiplier. The voltage generated in the photomultiplier was used to drive galvanometers which had a small mirror mounted on the axis. The moving reflected light beam would strike the surface of the turning drum, which was covered with photo-sensitive paper. The expense of developing photo sensitive paper caused many seismic observatories to switch to ink or thermal-sensitive paper.

Modern Instruments

Modern instruments use electronic sensors, amplifiers, and recording devices. Most are broadband covering a wide range of frequencies. Some seismometers can measure motions with frequencies from 500 Hz to 0.00118 Hz (1/500 = 0.002 seconds per cycle, to 1/0.00118 = 850 seconds per cycle). The mechanical suspension for horizontal instruments remains the garden-gate described above. Vertical instruments use some kind of constant-force suspension, such as the LaCoste suspension. The LaCoste suspension uses a zero-length spring to provide a long period (high sensitivity). Some modern instruments use a "triaxial" design, in which three identical motion sensors are set at the same angle to the vertical but 120 degrees apart on the horizontal. Vertical and horizontal motions can be computed from the outputs of the three sensors.

CMG-40T triaxial broadband seismometer

Seismometers unavoidably introduce some distortion into the signals they measure, but professionally designed systems have carefully characterized frequency transforms.

Modern sensitivities come in three broad ranges: geophones, 50 to 750 V/m; local geologic seismographs, about 1,500 V/m; and teleseismographs, used for world survey, about 20,000 V/m. Instruments come in three main varieties: short period, long period and broadband. The short and long period measure velocity and are very sensitive, however they 'clip' the signal or go off-scale for ground motion that is strong enough to be felt by people. A 24-bit analog-to-digital conversion channel is commonplace. Practical devices are linear to roughly one part per million.

Delivered seismometers come with two styles of output: analog and digital. Analog seismographs require analog recording equipment, possibly including an analog-to-digital converter. The output of a digital seismograph can be simply input to a computer. It presents the data in a standard digital format (often "SE2" over Ethernet).

Teleseismometers

The modern broadband seismograph can record a very broad range of frequencies. It consists of a small "proof mass", confined by electrical forces, driven by sophisticated electronics. As the earth moves, the electronics attempt to hold the mass steady through a feedback circuit. The amount of force necessary to achieve this is then recorded.

A low-frequency 3-direction ocean-bottom seismometer (cover removed). Two masses for x- and y-direction can be seen, the third one for z-direction is below. This model is a CMG-40TOBS, manufactured by Güralp Systems Ltd and is part of the Monterey Accelerated Research System.

In most designs the electronics holds a mass motionless relative to the frame. This device is called a "force balance accelerometer". It measures acceleration instead of velocity of ground movement. Basically, the distance between the mass and some part of the frame is measured very precisely, by a linear variable differential transformer. Some instruments use a linear variable differential capacitor.

That measurement is then amplified by electronic amplifiers attached to parts of an electronic negative feedback loop. One of the amplified currents from the negative feedback loop drives a coil very like a loudspeaker, except that the coil is attached to the mass, and the magnet is mounted on the frame. The result is that the mass stays nearly motionless.

Most instruments measure directly the ground motion using the distance sensor. The voltage generated in a sense coil on the mass by the magnet directly measures the instantaneous velocity of the ground. The current to the drive coil provides a sensitive, accurate measurement of the force between the mass and frame, thus measuring directly the ground's acceleration (using f=ma where f=force, m=mass, a=acceleration).

One of the continuing problems with sensitive vertical seismographs is the buoyancy of their masses. The uneven changes in pressure caused by wind blowing on an open window can easily change the density of the air in a room enough to cause a vertical seismograph to show spurious signals. Therefore, most professional seismographs are sealed in rigid gas-tight enclosures. For example, this is why a common Streckeisen model has a thick glass base that must be glued to its pier without bubbles in the glue.

It might seem logical to make the heavy magnet serve as a mass, but that subjects the seismograph to errors when the Earth's magnetic field moves. This is also why seismograph's moving parts are constructed from a material that interacts minimally with magnetic fields. A seismograph is also sensitive to changes in temperature so many instruments are constructed from low expansion materials such as nonmagnetic invar.

The hinges on a seismograph are usually patented, and by the time the patent has expired, the design has been improved. The most successful public domain designs use thin foil hinges in a clamp.

Another issue is that the transfer function of a seismograph must be accurately characterized, so that its frequency response is known. This is often the crucial difference between professional and amateur instruments. Most instruments are characterized on a variable frequency shaking table.

Strong-motion Seismometers

Another type of seismometer is a digital strong-motion seismometer, or accelerograph. The data from such an instrument is essential to understand how an earthquake affects manmade structures.

A strong-motion seismometer measures acceleration. This can be mathematically integrated later to give velocity and position. Strong-motion seismometers are not as sensitive to ground motions as teleseismic instruments but they stay on scale during the strongest seismic shaking.

Other Forms

A Kinemetrics seismograph, formerly used by the United States Department of the Interior.

Accelerographs and geophones are often heavy cylindrical magnets with a spring-mounted coil inside. As case moves, the coil tends to stay stationary, so the magnetic field cuts the wires, inducing current in the output wires. They receive frequencies from several hundred hertz down to 1 Hz. Some have electronic damping, a low-budget way to get some of the performance of the closed-loop wide-band geologic seismographs.

Strain-beam accelerometers constructed as integrated circuits are too insensitive for geologic seismographs (2002), but are widely used in geophones.

Some other sensitive designs measure the current generated by the flow of a non-corrosive ionic fluid through an electret sponge or a conductive fluid through a magnetic field.

Interconnected Seismometers

Seismometers spaced in an array can also be used to precisely locate, in three dimensions, the source of an earthquake, using the time it takes for seismic waves to propagate away from the hypocenter, the initiating point of fault rupture. Interconnected seismometers are also used to detect underground nuclear test explosions, as well as for Earthquake early warning systems. These seismometer are often used as part of a large scale governmental or scientific project, but some organizations such as the Quake-Catcher Network, can use residential size detectors built into computers to detect earthquakes as well.

In reflection seismology, an array of seismometers image sub-surface features. The data are reduced to images using algorithms similar to tomography. The data reduction methods resemble those of computer-aided tomographic medical imaging X-ray machines (CAT-scans), or imaging sonars.

A worldwide array of seismometers can actually image the interior of the Earth in wave-speed and transmissivity. This type of system uses events such as earthquakes, impact events or nuclear explosions as wave sources. The first efforts at this method used manual data reduction from paper seismograph charts. Modern digital seismograph records are better adapted to direct computer use. With inexpensive seismometer designs and internet access, amateurs and small institutions have even formed a "public seismograph network."

Seismographic systems used for petroleum or other mineral exploration historically used an explosive and a wireline of geophones unrolled behind a truck. Now most short-range systems use "thumpers" that hit the ground, and some small commercial systems have such good digital signal processing that a few sledgehammer strikes provide enough signal for short-distance refractive surveys. Exotic cross or two-dimensional arrays of geophones are sometimes used to perform three-dimensional reflective imaging of subsurface features. Basic linear refractive geomapping software (once a black art) is available off-the-shelf, running on laptop computers, using strings as small as three geophones. Some systems now come in an 18" (0.5 m) plastic field case with a computer, display and printer in the cover.

Small seismic imaging systems are now sufficiently inexpensive to be used by civil engineers to survey foundation sites, locate bedrock, and find subsurface water.

Recording

Today, the most common recorder is a computer with an analog-to-digital converter, a disk drive and an internet connection; for amateurs, a PC with a sound card and associated software is adequate. Most systems record continuously, but some record only when a signal is detected, as shown by a short-term increase in the variation of the signal, compared to its long-term average (which can vary slowly because of changes in seismic noise).

Viewing of a Develocorder film

Matsushiro Seismological Observatory

Prior to the availability of digital processing of seismic data in the late 1970s, the records were done in a few different forms on different types of media. A "Helicorder" drum was a device used to record data into photographic paper or in the form of paper and ink. A "Develocorder" was a machine that record data from up to 20 channels into a 16-mm film. The recorded film can be viewed by a machine. The reading and measuring from these types of media can be done by hand. After the digital processing has been used, the archives of the seismic data were recorded in magnetic tapes. Due to the deterioration of older magnetic tape medias, large number of waveforms from the archives are not recoverable.

References

- Craig Taylor and Erik VanMarcke, ed. (2002). Acceptable Risk Processes: Lifeline and Natural Hazards. Reston, VA: ASCE, TCLEE. ISBN 9780784406236.

- Reitherman, Robert (2012). Earthquakes and Engineers: An International History. Reston, VA:

ASCE Press. pp. 486–487. ISBN 9780784410714.

- Ben-Menahem, A. (2009). Historical Encyclopedia of Natural and Mathematical Sciences , Volume 1. Springer. p. 2657. ISBN 9783540688310. Retrieved 28 August 2012.

- William H.K. Lee; Paul Jennings; Carl Kisslinger; Hiroo Kanamori (27 September 2002). International Handbook of Earthquake & Engineering Seismology. Academic Press. pp. 283–. ISBN 978-0-08-048922-3. Retrieved 29 April 2013.

- Reitherman, Robert (2012). Earthquakes and Engineers: an International History. Reston, VA: ASCE Press. pp. 122–125. ISBN 9780784410714.

- Finazzi, Francesco (2016). "The Earthquake Network Project: Toward a Crowdsourced Smartphone - Based Earthquake Early Warning System". Bulletin of the Seismological Society of America. doi:10.1785/0120150354. Retrieved 10 June 2016.

- Verrucci E et al (2016) Digital engagement methods for earthquake and fire preparedness—a review. Nat Hazards 83:1583

- Lin, Rong-gong (2015-03-25). "Congress members urge $16 million to fund quake early warning system". Los Angeles Times. Retrieved 2015-10-24.

- "USGS Release: USGS Awards $4 Million to Support Earthquake Early Warning System in California and Pacific Northwest (7/30/2015 12:00:00 PM)". www.usgs.gov. Retrieved 2015-08-23.

- Jardin, Xeni. "Earthquake early warning system gets a $4 million boost from USGS". Boing Boing. Retrieved 2015-08-23.

- "Hospital Evacuation Decision Guide: Chapter 2. Pre-Disaster Self-Assessment". Archive.ahrq. gov. 2011-06-30. Retrieved 2015-03-08.

- Lin II, Rong-Gong (December 14, 2014). "California receives U.S. funding for earthquake early-warning system". Los Angeles Times. Retrieved 2014-12-31.

- "Experimental warning system gave 10-second alert before California earthquake". CBS News. 2014-08-24. Archived from the original on 25 August 2014.

- Hutton, Kate; Yu, Ellen. "NEWS FLASH!! SCSN Earthquake Catalog Completed!!" (PDF). Seismological Laboratory, Caltech. Retrieved 4 July 2014.

- Smith, Charles (2006-04-15). "What San Francisco didn't learn from the '06 quake". San Francisco Chronicle. Retrieved 20 June 2011.

Earthquake Engineering: An Integrated Study

Earthquake engineering is a branch of engineering that studies buildings and structures in order to prevent them from earthquakes. The basic aim of this subject is to make buildings resistant to earthquakes. This text helps the reader in developing an in-depth understanding of the subject matter.

Earthquake Engineering

Earthquake engineering is an interdisciplinary branch of engineering that designs and analyzes structures, such as buildings and bridges, with earthquakes in mind. Its overall goal is to make such structures more resistant to earthquakes. An earthquake (or seismic) engineer aims to construct structures that will not be damaged in minor shaking and will avoid serious damage or collapse in a major earthquake. Earthquake engineering is the scientific field concerned with protecting society, the natural environment, and the man-made environment from earthquakes by limiting the seismic risk to socio-economically acceptable levels. Traditionally, it has been narrowly defined as the study of the behavior of structures and geo-structures subject to seismic loading; it is considered as a subset of structural engineering, geotechnical engineering, mechanical engineering, chemical engineering, applied physics, etc. However, the tremendous costs experienced in recent earthquakes have led to an expansion of its scope to encompass disciplines from the wider field of civil engineering, mechanical engineering and from the social sciences, especially sociology, political science, economics and finance.

The main objectives of earthquake engineering are:

- Foresee the potential consequences of strong earthquakes on urban areas and civil infrastructure.

- Design, construct and maintain structures to perform at earthquake exposure up to the expectations and in compliance with building codes.

A properly engineered structure does not necessarily have to be extremely strong or expensive. It has to be properly designed to withstand the seismic effects while sustaining an acceptable level of damage.

Seismic Loading

Seismic loading means application of an earthquake-generated excitation on a structure (or geo-structure). It happens at contact surfaces of a structure either with the ground, with adjacent structures, or with gravity waves from tsunami. The loading that is expected at a given location on the Earth's surface is estimated by engineering seismology. It is related to the seismic hazard of the location.

Tokyo Skytree, equipped with a tuned mass damper, is the world's tallest tower and is the world's second tallest structure.

Seismic Performance

Earthquake or seismic performance defines a structure's ability to sustain its main functions, such as its safety and serviceability, *at* and *after* a particular earthquake exposure. A structure is normally considered *safe* if it does not endanger the lives and well-being of those in or around it by partially or completely collapsing. A structure may be considered *serviceable* if it is able to fulfill its operational functions for which it was designed.

Basic concepts of the earthquake engineering, implemented in the major building codes, assume that a building should survive a rare, very severe earthquake by sustaining significant damage but without globally collapsing. On the other hand, it should remain operational for more frequent, but less severe seismic events.

Seismic Performance Assessment

Engineers need to know the quantified level of the actual or anticipated seismic performance associated with the direct damage to an individual building subject to a specified ground shaking. Such an assessment may be performed either experimentally or analytically.

Experimental Assessment

Experimental evaluations are expensive tests that are typically done by placing a (scaled) model of the structure on a shake-table that simulates the earth shaking and observing its behavior. Such kinds of experiments were first performed more than a century ago. Only recently has it become possible to perform 1:1 scale testing on full structures.

Due to the costly nature of such tests, they tend to be used mainly for understanding the seismic behavior of structures, validating models and verifying analysis methods. Thus, once properly validated, computational models and numerical procedures tend to carry the major burden for the seismic performance assessment of structures.

Analytical/Numerical Assessment

Seismic performance assessment or seismic structural analysis is a powerful tool of earthquake engineering which utilizes detailed modelling of the structure together with methods of structural analysis to gain a better understanding of seismic performance of building and non-building structures. The technique as a formal concept is a relatively recent development.

In general, seismic structural analysis is based on the methods of structural dynamics. For decades, the most prominent instrument of seismic analysis has been the earthquake response spectrum method which also contributed to the proposed building code's concept of today.

However, such methods are good only for linear elastic systems, being largely unable to model the structural behavior when damage (i.e., non-linearity) appears. Numerical *step-by-step integration* proved to be a more effective method of analysis for multi-degree-of-freedom structural systems with significant non-linearity under a transient process of ground motion excitation.

Basically, numerical analysis is conducted in order to evaluate the seismic performance of buildings. Performance evaluations are generally carried out by using nonlinear static pushover analysis or nonlinear time-history analysis. In such analyses, it is essential to achieve accurate non-linear modeling of structural components such as beams, columns, beam-column joints, shear walls etc. Thus, experimental results play an important role in determining the modeling parameters of individual components, especially those that are subject to significant non-linear deformations. The individual components are then assembled to create a full non-linear model of the structure. Thus created models are analyzed to evaluate the performance of buildings.

The capabilities of the structural analysis software are a major consideration in the above process as they restrict the possible component models, the analysis methods available and, most importantly, the numerical robustness. The latter becomes a major consideration for structures that venture into the non-linear range and approach global or local collapse as the numerical solution becomes increasingly unstable and thus difficult to reach. There are several commercially available Finite Element Analysis software's such as CSI-SAP2000 and CSI-PERFORM-3D and Scia Engineer-ECtools which can be used for the seismic performance evaluation of buildings. Moreover, there is research-based finite element analysis platforms such as OpenSees, RUAUMOKO and the older DRAIN-2D/3D, several of which are now open source.

Research for Earthquake Engineering

Research for earthquake engineering means both field and analytical investigation or experimentation intended for discovery and scientific explanation of earthquake engineering related facts, revision of conventional concepts in the light of new findings, and practical application of the developed theories.

The National Science Foundation (NSF) is the main United States government agency that supports fundamental research and education in all fields of earthquake engineering. In particular, it focuses on experimental, analytical and computational research on design and performance enhancement of structural systems.

E-Defense Shake Table

The Earthquake Engineering Research Institute (EERI) is a leader in dissemination of earthquake engineering research related information both in the U.S. and globally.

A definitive list of earthquake engineering research related shaking tables around the world may be found in Experimental Facilities for Earthquake Engineering Simulation Worldwide. The most prominent of them is now E-Defense Shake Table in Japan.

Major U.S. Research Programs

NSF also supports the George E. Brown, Jr. Network for Earthquake Engineering Simulation

The NSF Hazard Mitigation and Structural Engineering program (HMSE) supports research on new technologies for improving the behavior and response of structural systems subject to earthquake hazards; fundamental research on safety and reliability of constructed systems; innovative developments in analysis and model based simulation of structural behavior and response including soil-structure interaction; design concepts that improve structure performance and flexibility; and application of new control techniques for structural systems.

(NEES) that advances knowledge discovery and innovation for earthquakes and tsunami loss reduction of the nation's civil infrastructure and new experimental simulation techniques and instrumentation.

The NEES network features 14 geographically-distributed, shared-use laboratories that support several types of experimental work: geotechnical centrifuge research, shake-table tests, large-scale structural testing, tsunami wave basin experiments, and field site research. Participating universities include: Cornell University; Lehigh University; Oregon State University; Rensselaer Polytechnic Institute; University at Buffalo, State University of New York; University of California, Berkeley; University of California, Davis; University of California, Los Angeles; University of California, San Diego; University of California, Santa Barbara; University of Illinois, Urbana-Champaign; University of Minnesota; University of Nevada, Reno; and the University of Texas, Austin.

NEES at Buffalo testing facility

The equipment sites (labs) and a central data repository are connected to the global earthquake engineering community via the NEEShub website. The NEES website is powered by HUBzero software developed at Purdue University for nanoHUB specifically to help the scientific community share resources and collaborate. The cyberinfrastructure, connected via Internet2, provides interactive simulation tools, a simulation tool development area, a curated central data repository, animated presentations, user support, telepresence, mechanism for uploading and sharing resources, and statistics about users and usage patterns.

This cyberinfrastructure allows researchers to: securely store, organize and share data within a standardized framework in a central location; remotely observe and participate in experiments through the use of synchronized real-time data and video; collaborate with colleagues to facilitate the planning, performance, analysis, and publication of research experiments; and conduct computational and hybrid simulations that may combine the results of multiple distributed experiments and link physical experiments with computer simulations to enable the investigation of overall system performance.

These resources jointly provide the means for collaboration and discovery to improve the seismic design and performance of civil and mechanical infrastructure systems.

Earthquake Simulation

The very first earthquake simulations were performed by statically applying some *horizontal inertia forces* based on scaled peak ground accelerations to a mathematical model of a building. With the further development of computational technologies, static approaches began to give way to dynamic ones.

Dynamic experiments on building and non-building structures may be physical, like shake-table testing, or virtual ones. In both cases, to verify a structure's expected seismic performance, some researchers prefer to deal with so called "real time-histories" though the last cannot be "real" for a hypothetical earthquake specified by either a building code or by some particular research requirements. Therefore, there is a strong incentive to engage an earthquake simulation which is the seismic input that possesses only essential features of a real event.

Sometimes earthquake simulation is understood as a re-creation of local effects of a strong earth shaking.

Structure Simulation

Theoretical or experimental evaluation of anticipated seismic performance mostly requires a structure simulation which is based on the concept of structural likeness or similarity. Similarity is some degree of analogy or resemblance between two or more objects. The notion of similarity rests either on exact or approximate repetitions of patterns in the compared items.

In general, a building model is said to have similarity with the real object if the two share *geometric similarity, kinematic similarity* and *dynamic similarity*. The most vivid and effective type of similarity is the *kinematic* one. *Kinematic similarity* exists when the paths and velocities of moving particles of a model and its prototype are similar.

The ultimate level of *kinematic similarity* is *kinematic equivalence* when, in the case of earthquake engineering, time-histories of each story lateral displacements of the model and its prototype would be the same.

Seismic Vibration Control

Seismic vibration control is a set of technical means aimed to mitigate seismic impacts in building and non-building structures. All seismic vibration control devices may be classified as *passive, active* or *hybrid* where:

- *passive control devices* have no feedback capability between them, structural elements and the ground;

- *active control devices* incorporate real-time recording instrumentation on the ground integrated with earthquake input processing equipment and actuators within the structure;

- *hybrid control devices* have combined features of active and passive control systems.

When ground seismic waves reach up and start to penetrate a base of a building, their energy flow density, due to reflections, reduces dramatically: usually, up to 90%. However, the remaining portions of the incident waves during a major earthquake still bear a huge devastating potential.

After the seismic waves enter a superstructure, there are a number of ways to control them in order to soothe their damaging effect and improve the building's seismic performance, for instance:

- to dissipate the wave energy inside a superstructure with properly engineered dampers;

- to disperse the wave energy between a wider range of frequencies;

- to absorb the resonant portions of the whole wave frequencies band with the help of so-called *mass dampers.*

Mausoleum of Cyrus, the oldest base-isolated structure in the world

Devices of the last kind, abbreviated correspondingly as TMD for the tuned (*passive*), as AMD for the *active*, and as HMD for the *hybrid mass dampers*, have been studied and installed in high-rise buildings, predominantly in Japan, for a quarter of a century.

However, there is quite another approach: partial suppression of the seismic energy flow into the superstructure known as seismic or base isolation.

For this, some pads are inserted into or under all major load-carrying elements in the base of the building which should substantially decouple a superstructure from its sub-structure resting on a shaking ground.

The first evidence of earthquake protection by using the principle of base isolation was dis-covered in Pasargadae, a city in ancient Persia, now Iran, and dates back to the 6th century BCE. Below, there are some samples of seismic vibration control technologies of today.

Dry-stone Walls Control

Dry-stone walls of Machu Picchu Temple of the Sun, Peru

People of Inca civilization were masters of the polished *'dry-stone walls'*, called ashlar, where blocks of stone were cut to fit together tightly without any mortar. The Incas were among the best stonemasons the world has ever seen and many junctions in their masonry were so perfect that even blades of grass could not fit between the stones.

Peru is a highly seismic land and for centuries the mortar-free construction proved to be apparently more earthquake-resistant than using mortar. The stones of the dry-stone walls built by the Incas could move slightly and resettle without the walls col-lapsing, a passive structural control technique employing both the principle of energy dissipation and that of suppressing resonant amplifications.

Tuned Mass Damper

Typically the tuned mass dampers are huge concrete blocks mounted in skyscrapers or other structures and moved in opposition to the resonance frequency oscillations of the structures by means of some sort of spring mechanism.

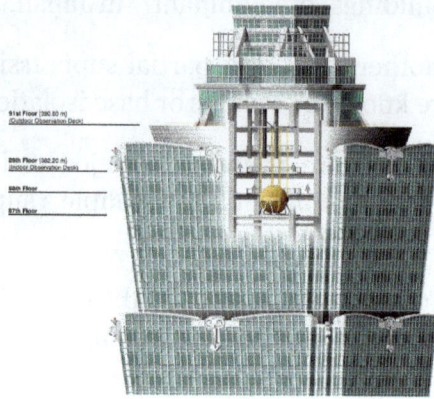

Tuned mass damper in Taipei 101, the world's third tallest skyscraper

Taipei 101 skyscraper needs to withstand typhoon winds and earthquake tremors common in its area of the Asia-Pacific. For this purpose, a steel pendulum weighing 660 metric tones that serves as a tuned mass damper was designed and installed atop the structure. Suspended from the 92nd to the 88th floor, the pendulum sways to decrease resonant amplifications of lateral displacements in the building caused by earthquakes and strong gusts.

Building Elevation Control

Transamerica Pyramid building

Building elevation control is a valuable source of vibration control of seismic loading. Pyramid-shaped skyscrapers continue to attract the attention of architects and engineers because such structures promise a better stability against earthquakes and winds. The elevation configuration can prevent buildings' resonant amplifications because a properly configured building disperses the shear wave energy between a wide range of frequencies.

Earthquake or wind quieting ability of the elevation configuration is provided by a specific pattern of multiple reflections and transmissions of vertically propagating waves, which are generated by breakdowns into homogeneity of story layers, and a taper. Any abrupt changes of the propagating waves velocity result in a considerable dispersion of the wave energy between a wide ranges of frequencies thus preventing the resonant displacement amplifications in the building.

A tapered profile of a building is not a compulsory feature of this method of structural control. A similar resonance preventing effect can be also obtained by a proper *tapering* of other characteristics of a building structure, namely, its mass and stiffness. As a result, the building elevation configuration techniques permit an architectural design that may be both attractive and functional.

Hysteretic Dampers

Hysteretic damper is intended to provide better and more reliable seismic performance than that of a conventional structure at the expense of the seismic input energy dissipation. There are five major groups of hysteretic dampers used for the purpose, namely:

- Fluid viscous dampers (FVDs)

Viscous Dampers have the benefit of being a supplemental damping system. They have an oval hysteretic loop and the damping is velocity dependent. While some minor maintenance is potentially required, viscous dampers generally do not need to be replaced after an earthquake. While more expensive than other damping technologies they can be used for both seismic and wind loads and are the most commonly used hysteric damper.

- Friction dampers (FDs)

Friction dampers tend to be available in two major types, linear and rotational and dissipate energy by heat. The damper operates on the principle of a coulomb damper. Depending on the design, friction dampers can experience stick-slip phenomenon and Cold welding. The main disadvantage being that friction surfaces can wear over time and for this reason they are not recommended for dissipating wind loads. When used in seismic applications wear is not a problem and there is no required maintenance. They have a rectangular hysteretic loop and as long as the building is sufficiently elastic they tend to settle back to their original positions after an earthquake.

- Metallic yielding dampers (MYDs)

Metallic yielding dampers, as the name implies, yield in order to absorb the earthquake's energy. This type of damper absorbs a large amount of energy however they must be replaced after an earthquake and may prevent the building from settling back to its original position.

- Viscoelastic dampers (VEDs)

Viscoelastic dampers are useful in that they can be used for both wind and seismic applications, they are usually limited to small displacements. There is some concern as to the reliability of the technology as some brands have been banned from use in buildings in the United States.

- Straddlingpendulum dampers (swing)

Base Isolation

Base Isolation seeks to prevent the kinetic energy of the earthquake from being transferred into elastic energy in the building. These technologies do so by isolating the structure from the ground, thus enabling them to move somewhat independently. The degree to which the energy is transferred into the strucuture and how the energy is dissipated will vary depending on the technology used.

- Lead Rubber Bearing

Lead Rubber Bearing or LRB is a type of base isolation employing a heavy damping. It was invented by Bill Robinson, a New Zealander.

Heavy damping mechanism incorporated in vibration control technologies and, particularly, in base isolation devices, is often considered a valuable source of suppressing vibrations thus enhancing a building's seismic performance. However, for the rather pliant systems such as base isolated structures, with a relatively low bearing stiffness but with a high damping, the so-called "damping force" may turn out the main pushing force at a strong earthquake. The video shows a Lead Rubber Bearing being tested at the UCSD Caltrans-SRMD facility. The bearing is made of rubber with a lead core. It was a uniaxial test in which the bearing was also under a full structure load. Many buildings and bridges, both in New Zealand and elsewhere, are protected with lead dampers and lead and rubber bearings. Te Papa Tongarewa, the national museum of New Zealand, and the New Zealand Parliament Buildings have been fitted with the bearings. Both are in Wellington which sits on an active earthquake fault.

- Springs-with-damper base isolator

Springs-with-damper base isolator installed under a three-story town-house, Santa Monica, California is shown on the photo taken prior to the 1994 Northridge earthquake exposure. It is a base isolation device conceptually similar to *Lead Rubber Bearing.*

Springs-with-damper close-up

One of two three-story town-houses like this, which was well instrumented for recording of both vertical and horizontal accelerations on its floors and the ground, has survived a severe shaking during the Northridge earthquake and left valuable recorded information for further study.

- Simple roller bearing

Simple roller bearing is a base isolation device which is intended for protection of various building and non-building structures against potentially damaging lateral impacts of strong earthquakes.

This metallic bearing support may be adapted, with certain precautions, as a seismic isolator to skyscrapers and buildings on soft ground. Recently, it has been employed under the name of Metallic Roller Bearing for a housing complex (17 stories) in Tokyo, Japan.

- Friction pendulum bearing

FPB shake-table testing

Friction Pendulum Bearing (FPB) is another name of Friction Pendulum System (FPS). It is based on three pillars:

- articulated friction slider;

- spherical concave sliding surface;

- enclosing cylinder for lateral displacement restraint.

Snapshot with the link to video clip of a shake-table testing of FPB system supporting a rigid building model is presented at the right.

Seismic Design

Seismic design is based on authorized engineering procedures, principles and criteria meant to design or retrofit structures subject to earthquake exposure. Those criteria are only consistent with the contemporary state of the knowledge about earthquake engineering structures. Therefore, a building design which exactly follows seismic code regulations does not guarantee safety against collapse or serious damage.

The price of poor seismic design may be enormous. Nevertheless, seismic design has always been a trial and error process whether it was based on physical laws or on empirical knowledge of the structural performance of different shapes and materials.

San Francisco City Hall destroyed by 1906 earthquake and fire.

San Francisco after the 1906 earthquake and fire

To practice seismic design, seismic analysis or seismic evaluation of new and existing civil engineering projects, an engineer should, normally, pass examination on *Seismic Principles* which, in the State of California, include:

- Seismic Data and Seismic Design Criteria

- Seismic Characteristics of Engineered Systems

- Seismic Forces

- Seismic Analysis Procedures

- Seismic Detailing and Construction Quality Control

To build up complex structural systems, seismic design largely uses the same relatively small number of basic structural elements (to say nothing of vibration control devices) as any non-seismic design project.

Normally, according to building codes, structures are designed to "withstand" the largest earthquake of a certain probability that is likely to occur at their location. This means the loss of life should be minimized by preventing collapse of the buildings.

Seismic design is carried out by understanding the possible failure modes of a structure and providing the structure with appropriate strength, stiffness, ductility, and configuration to ensure those modes cannot occur.

Seismic Design Requirements

Seismic design requirements depend on the type of the structure, locality of the project and its authorities which stipulate applicable seismic design codes and criteria. For instance, California Department of Transportation's requirements called *The Seismic Design Criteria* (SDC) and aimed at the design of new bridges in California incorporate an innovative seismic performance-based approach.

The Metsamor Nuclear Power Plant was closed after the 1988 Armenian earthquake

The most significant feature in the SDC design philosophy is a shift from a *force-based assessment* of seismic demand to a *displacement-based assessment* of demand and capacity. Thus, the newly adopted displacement approach is based on comparing the *elastic displacement* demand to the *inelastic displacement* capacity of the primary structural components while ensuring a minimum level of inelastic capacity at all potential plastic hinge locations.

In addition to the designed structure itself, seismic design requirements may include a *ground stabilization* underneath the structure: sometimes, heavily shaken ground breaks up which leads to collapse of the structure sitting upon it. The following topics should be of primary concerns: liquefaction; dynamic lateral earth pressures on retaining walls; seismic slope stability; earthquake-induced settlement.

Nuclear facilities should not jeopardise their safety in case of earthquakes or other hostile external events. Therefore, their seismic design is based on criteria far more stringent than those applying to non-nuclear facilities. The Fukushima I nuclear accidents and damage to other nuclear facilities that followed the 2011 Tōhoku earthquake and tsunami have, however, drawn attention to ongoing concerns over Japanese nuclear seismic design standards and caused other many governments to re-evaluate their nuclear programs. Doubt has also been expressed over the seismic evaluation and design of certain other plants, including the Fessenheim Nuclear Power Plant in France.

Failure Modes

Failure mode is the manner by which an earthquake induced failure is observed. It, generally, describes the way the failure occurs. Though costly and time consuming, learning from each real earthquake failure remains a routine recipe for advancement in *seismic design* methods. Below, some typical modes of earthquake-generated failures are presented.

Typical damage to unreinforced masonry buildings at earthquakes

The lack of reinforcement coupled with poor mortar and inadequate roof-to-wall ties can result in substantial damage to an unreinforced masonry building. Severely cracked or leaning walls are some of the most common earthquake damage. Also hazardous is the damage that may occur between the walls and roof or floor diaphragms. Separation between the framing and the walls can jeopardize the vertical support of roof and floor systems.

Soft story collapse due to inadequate shear strength at ground level, Loma Prieta earthquake

Soft story effect. Absence of adequate shear walls on the ground level caused damage to this structure. A close examination of the image reveals that the rough board siding, once covered by a brick veneer, has been completely dismantled from the studwall. Only the rigidity of the floor above combined with the support on the two hidden sides by continuous walls, not penetrated with large doors as on the street sides, is preventing full collapse of the structure.

Soil liquefaction. In the cases where the soil consists of loose granular deposited materials with the tendency to develop excessive hydrostatic pore water pressure of sufficient magnitude and compact, liquefaction of those loose saturated deposits may result in non-uniform settlements and tilting of structures. This caused major damage to thousands of buildings in Niigata, Japan during the 1964 earthquake.

Car smashed by landslide rock, 2008 Sichuan earthquake

Landslide rock fall. A landslide is a geological phenomenon which includes a wide range of ground movement, including rock falls. Typically, the action of gravity is the primary driving force for a landslide to occur though in this case there was another contributing factor which affected the original slope stability: the landslide required an *earthquake trigger* before being released.

Pounding against adjacent building. This is a photograph of the collapsed five-story tower, St. Joseph's Seminary, Los Altos, California which resulted in one fatality. During Loma Prieta earthquake, the tower pounded against the independently vibrating adjacent building behind. A possibility of pounding depends on both buildings' lateral displacements which should be accurately estimated and accounted for.

Effects of completely shattered joints of concrete frame, Northridge

At Northridge earthquake, the Kaiser Permanente concrete frame office building had joints completely shattered, revealing inadequate confinement steel, which resulted in the second story collapse. In the transverse direction, composite end shear walls, consisting of two wythes of brick and a layer of shotcrete that carried the lateral load, peeled apart because of inadequate through-ties and failed.

7-story reinforced concrete buildings on steep slope collapse due to the following:

- Improper construction site on a foothill.

- Poor detailing of the reinforcement (lack of concrete confinement in the columns and at the beam-column joints, inadequate splice length).

- Seismically weak soft story at the first floor.

- Long cantilevers with heavy dead load.

shifting from foundation, Whittier

Sliding off foundations effect of a relatively rigid residential building structure during 1987 Whittier Narrows earthquake. The magnitude 5.9 earthquake pounded the Garvey West Apartment building in Monterey Park, California and shifted its superstructure about 10 inches to the east on its foundation.

Earthquake damage in Pichilemu.

If a superstructure is not mounted on a base isolation system, its shifting on the basement should be prevented.

Insufficient shear reinforcement let main rebars to buckle, Northridge

Reinforced concrete column burst at Northridge earthquake due to insufficient shear reinforcement mode which allows main reinforcement to buckle outwards. The deck unseated at the hinge and failed in shear. As a result, the La Cienega-Venice underpass section of the 10 Freeway collapsed.

Support-columns and upper deck failure, Loma Prieta earthquake

Loma Prieta earthquake: side view of reinforced concrete support-columns failure which triggered the upper deck collapse onto the lower deck of the two-level Cypress viaduct of Interstate Highway 880, Oakland, CA.

Failure of retaining wall due to ground movement, Loma Prieta

Retaining wall failure at Loma Prieta earthquake in Santa Cruz Mountains area: prominent northwest-trending extensional cracks up to 12 cm (4.7 in) wide in the concrete spillway to Austrian Dam, the north abutment.

Lateral spreading mode of ground failure, Loma Prieta

Ground shaking triggered soil liquefaction in a subsurface layer of sand, producing differential lateral and vertical movement in an overlying carapace of unliquified sand and silt. This mode of ground failure, termed lateral spreading, is a principal cause of liquefaction-related earthquake damage.

Beams and pier columns diagonal cracking, 2008 Sichuan earthquake

Severely damaged building of Agriculture Development Bank of China after 2008 Sichuan earthquake: most of the beams and pier columns are sheared. Large diagonal cracks in masonry and veneer are due to in-plane loads while abrupt settlement of the right end of the building should be attributed to a landfill which may be hazardous even without any earthquake.

Twofold tsunami impact: sea waves hydraulic pressure and inundation. Thus, the Indian Ocean earthquake of December 26, 2004, with the epicenter off the west coast of Sumatra, Indonesia, triggered a series of devastating tsunamis, killing more than 230,000 people in eleven countries by inundating surrounding coastal communities with huge waves up to 30 meters (100 feet) high.

Earthquake-resistant Construction

Earthquake construction means implementation of seismic design to enable building and non-building structures to live through the anticipated earthquake exposure up to the expectations and in compliance with the applicable building codes.

Construction of Pearl River Tower X-bracing to resist lateral forces of earthquakes and winds

Design and construction are intimately related. To achieve a good workmanship, detailing of the members and their connections should be as simple as possible. As any construction in general, earthquake construction is a process that consists of the building, retrofitting or assembling of infrastructure given the construction materials available.

The destabilizing action of an earthquake on constructions may be *direct* (seismic motion of the ground) or *indirect* (earthquake-induced landslides, soil liquefaction and waves of tsunami).

A structure might have all the appearances of stability, yet offer nothing but danger when an earthquake occurs. The crucial fact is that, for safety, earthquake-resistant construction techniques are as important as quality control and using correct materials. *Earthquake contractor* should be registered in the state/provice/country of the project location (depending on local regulations), bonded and insured.

To minimize possible losses, construction process should be organized with keeping in mind that earthquake may strike any time prior to the end of construction.

Each construction project requires a qualified team of professionals who understand the basic features of seismic performance of different structures as well as construction management.

Adobe Structures

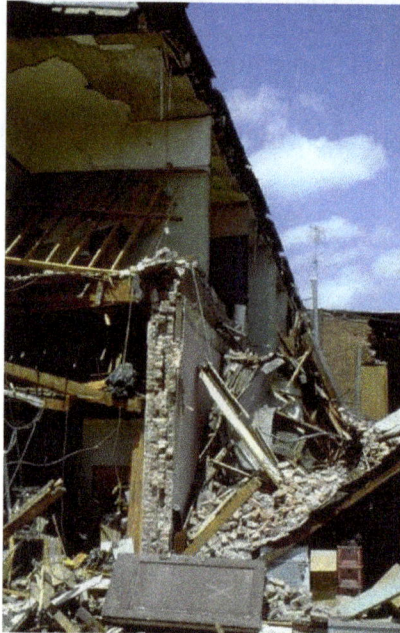

Partially collapsed adobe building in Westmorland, California

Around thirty percent of the world's population lives or works in earth-made construction. Adobe type of mud bricks is one of the oldest and most widely used building materials. The use of adobe is very common in some of the world's most hazard-prone regions, traditionally across Latin America, Africa, Indian subcontinent and other parts of Asia, Middle East and Southern Europe.

Adobe buildings are considered very vulnerable at strong quakes. However, multiple ways of seismic strengthening of new and existing adobe buildings are available.

Key factors for the improved seismic performance of adobe construction are:

- Quality of construction.
- Compact, box-type layout.
- Seismic reinforcement.

Limestone and Sandstone Structures

Limestone is very common in architecture, especially in North America and Europe. Many landmarks across the world are made of limestone. Many medieval churches and castles in Europe are made of limestone and sandstone masonry. They are the long-lasting materials but their rather heavy weight is not beneficial for adequate seismic performance.

Application of modern technology to seismic retrofitting can enhance the survivability of unreinforced masonry structures. As an example, from 1973 to 1989, the Salt Lake City and County Building in Utah was exhaustively renovated and repaired with an emphasis on preserving historical accuracy in appearance. This was done in concert with a seismic upgrade that placed the weak sandstone structure on base isolation foundation to better protect it from earthquake damage.

Timber Frame Structures

Anne Hvide's House, Denmark (1560)

Timber framing dates back thousands of years, and has been used in many parts of the world during various periods such as ancient Japan, Europe and medieval England in localities where timber was in good supply and building stone and the skills to work it were not.

The use of timber framing in buildings provides their complete skeletal framing which offers some structural benefits as the timber frame, if properly engineered, lends itself to better *seismic survivability*.

Light-frame Structures

Light-frame structures usually gain seismic resistance from rigid plywood shear walls and wood structural panel diaphragms. Special provisions for seismic load-resisting systems for all engineered wood structures requires consideration of diaphragm ratios, horizontal and vertical diaphragm shears, and connector/fastener values. In addition, collectors, or drag struts, to distribute shear along a diaphragm length are required.

Reinforced Masonry Structures

A construction system where steel reinforcement is embedded in the mortar joints of masonry or placed in holes and after filled with concrete or grout is called reinforced masonry.

The devastating 1933 Long Beach earthquake revealed that masonry construction should be improved immediately. Then, the California State Code made the reinforced masonry mandatory.

There are various practices and techniques to achieve reinforced masonry. The most common type is the reinforced hollow unit masonry. The effectiveness of both vertical and horizontal reinforcement strongly depends on the type and quality of the masonry, i.e. masonry units and mortar.

To achieve a ductile behavior of masonry, it is necessary that the shear strength of the wall is greater than the flexural strength.

Reinforced Concrete Structures

Stressed Ribbon pedestrian bridge over the Rogue River, Grants Pass, Oregon

Prestressed concrete cable-stayed bridge over Yangtze river

Reinforced concrete is concrete in which steel reinforcement bars (rebars) or fibers have been incorporated to strengthen a material that would otherwise be brittle. It can be used to produce beams, columns, floors or bridges.

Prestressed concrete is a kind of reinforced concrete used for overcoming concrete's natural weakness in tension. It can be applied to beams, floors or bridges with a longer span than is practical with ordinary reinforced concrete. Prestressing tendons (generally of high tensile steel cable or rods) are used to provide a clamping load which produces a compressive stress that offsets the tensile stress that the concrete compression member would, otherwise, experience due to a bending load.

To prevent catastrophic collapse in response earth shaking (in the interest of life safety), a traditional reinforced concrete frame should have ductile joints. Depending upon the methods used and the imposed seismic forces, such buildings may be immediately usable, require extensive repair, or may have to be demolished.

Prestressed Structures

Prestressed structure is the one whose overall integrity, stability and security depend, primarily, on a *prestressing*. *Prestressing* means the intentional creation of permanent stresses in a structure for the purpose of improving its performance under various service conditions.

Naturally pre-compressed exterior wall of Colosseum, Rome

There are the following basic types of prestressing:

- Pre-compression (mostly, with the own weight of a structure)

- Pretensioning with high-strength embedded tendons

- Post-tensioning with high-strength bonded or unbonded tendons

Today, the concept of prestressed structure is widely engaged in design of buildings, underground structures, TV towers, power stations, floating storage and offshore facilities, nuclear reactor vessels, and numerous kinds of bridge systems.

A beneficial idea of *prestressing* was, apparently, familiar to the ancient Rome archi-tects; look, e.g., at the tall attic wall of Colosseum working as a stabilizing device for the wall piers beneath.

Steel Structures

Steel structures are considered mostly earthquake resistant but some failures have occurred. A great number of welded steel moment-resisting frame buildings, which looked earthquake-proof, surprisingly experienced brittle behavior and were hazard-ously damaged in the 1994 Northridge earthquake. After that, the Federal Emergency Management Agency (FEMA) initiated development of repair techniques and new de-sign approaches to minimize damage to steel moment frame buildings in future earth-quakes.

Collapsed section of the San Francisco–Oakland Bay Bridge in response to Loma Prieta earthquake

For structural steel seismic design based on Load and Resistance Factor Design (LRFD) approach, it is very important to assess ability of a structure to develop and maintain its bearing resistance in the inelastic range. A measure of this ability is ductility, which may be observed in a *material itself*, in a *structural element*, or to a *whole structure*.

As a consequence of Northridge earthquake experience, the American Institute of Steel Construction has introduced AISC 358 "Pre-Qualified Connections for Special and in-termediate Steel Moment Frames." The AISC Seismic Design Provisions require that all Steel Moment Resisting Frames employ either connections contained in AISC 358, or the use of connections that have been subjected to pre-qualifying cyclic testing.

Prediction of Earthquake Losses

Earthquake loss estimation is usually defined as a *Damage Ratio* (DR) which is a ratio of the earthquake damage repair cost to the total value of a building. *Probable Maximum Loss* (PML) is a common term used for earthquake loss estimation, but it lacks a precise definition. In 1999, ASTM E2026 'Standard Guide for the Estimation of Building Damageability in Earthquakes' was produced in order to standardize the nomenclature for seismic loss estimation, as well as establish guidelines as to the review process and qualifications of the reviewer.

Earthquake loss estimations are also referred to as *Seismic Risk Assessments*. The risk assessment process generally involves determining the probability of various ground motions coupled with the vulnerability or damage of the building under those ground motions. The results are defined as a percent of building replacement value.

Seismic Analysis

Seismic analysis is a subset of structural analysis and is the calculation of the response of a building (or nonbuilding) structure to earthquakes. It is part of the process of structural design, earthquake engineering or structural assessment and retrofit in regions where earthquakes are prevalent.

As seen in the figure, a building has the potential to 'wave' back and forth during an earthquake (or even a severe wind storm). This is called the 'fundamental mode', and is the lowest frequency of building response. Most buildings, however, have higher modes of response, which are uniquely activated during earthquakes. The figure just shows the second mode, but there are higher 'shimmy' (abnormal vibration) modes. Nevertheless, the first and second modes tend to cause the most damage in most cases.

The earliest provisions for seismic resistance were the requirement to design for a lateral force equal to a proportion of the building weight (applied at each floor level). This approach was adopted in the appendix of the 1927 Uniform Building Code (UBC), which was used on the west coast of the United States. It later became clear that the dynamic properties of the structure affected the loads generated during an earthquake. In the Los Angeles County Building Code of 1943 a provision to vary the load based on the number of floor levels was adopted (based on research carried out at Caltech in collaboration with Stanford University and the U.S. Coast and Geodetic Survey, which started in 1937). The concept of "response spectra" was developed in the 1930s, but it wasn't until 1952 that a joint committee of the San Francisco Section of the ASCE and the Structural Engineers Association of Northern California (SEAONC) proposed using the building period (the inverse of the frequency) to determine lateral forces.

The University of California, Berkeley was an early base for computer-based seismic analysis of structures, led by Professor Ray Clough (who coined the term finite element). Students included Ed Wilson, who went on to write the program SAP in 1970, an early "Finite Element Analysis" program.

Earthquake engineering has developed a lot since the early days, and some of the more complex designs now use special earthquake protective elements either just in the foundation (base isolation) or distributed throughout the structure. Analyzing these types of structures requires specialized explicit finite element computer code, which divides time into very small slices and models the actual physics, much like common video games often have "physics engines". Very large and complex buildings can be modeled in this way (such as the Osaka International Convention Center).

Structural analysis methods can be divided into the following five categories.

Equivalent Static Analysis

This approach defines a series of forces acting on a building to represent the effect of earthquake ground motion, typically defined by a seismic design response spectrum. It assumes that the building responds in its fundamental mode. For this to be true, the building must be low-rise and must not twist significantly when the ground moves. The response is read from a design response spectrum, given the natural frequency of the building (either calculated or defined by the building code). The applicability of this method is extended in many building codes by applying factors to account for higher buildings with some higher modes, and for low levels of twisting. To account for effects due to "yielding" of the structure, many codes apply modification factors that reduce the design forces (e.g. force reduction factors).

Response Spectrum Analysis

This approach permits the multiple modes of response of a building to be taken into account (in the frequency domain). This is required in many building codes for all except very simple or very complex structures. The response of a structure can be defined as a combination of many special shapes (modes) that in a vibrating string correspond to the "harmonics". Computer analysis can be used to determine these modes for a structure. For each mode, a response is read from the design spectrum, based on the modal frequency and the modal mass, and they are then combined to provide an estimate of the total response of the structure. In this we have to calculate the magnitude of forces in all directions i.e. X, Y & Z and then see the effects on the building.. Combination methods include the following:

- absolute - peak values are added together

- square root of the sum of the squares (SRSS)

- complete quadratic combination (CQC) - a method that is an improvement on SRSS for closely spaced modes

The result of a response spectrum analysis using the response spectrum from a ground motion is typically different from that which would be calculated directly from a linear dynamic analysis using that ground motion directly, since phase information is lost in the process of generating the response spectrum.

In cases where structures are either too irregular, too tall or of significance to a community in disaster response, the response spectrum approach is no longer appropriate, and more complex analysis is often required, such as non-linear static analysis or dynamic analysis.

Linear Dynamic Analysis

Static procedures are appropriate when higher mode effects are not significant. This is generally true for short, regular buildings. Therefore, for tall buildings, buildings with torsional irregularities, or non-orthogonal systems, a dynamic procedure is required. In the linear dynamic procedure, the building is modelled as a multi-degree-of-freedom (MDOF) system with a linear elastic stiffness matrix and an equivalent viscous damping matrix.

The seismic input is modelled using either modal spectral analysis or time history analysis but in both cases, the corresponding internal forces and displacements are determined using linear elastic analysis. The advantage of these linear dynamic procedures with respect to linear static procedures is that higher modes can be considered. However, they are based on linear elastic response and hence the applicability decreases with increasing nonlinear behaviour, which is approximated by global force reduction factors.

In linear dynamic analysis, the response of the structure to ground motion is calculated in the time domain, and all phase information is therefore maintained. Only linear properties are assumed. The analytical method can use modal decomposition as a means of reducing the degrees of freedom in the analysis.

Nonlinear Static Analysis

In general, linear procedures are applicable when the structure is expected to remain nearly elastic for the level of ground motion or when the design results in nearly uniform distribution of nonlinear response throughout the structure. As the performance objective of the structure implies greater inelastic demands, the uncertainty with linear procedures increases to a point that requires a high level of conservatism in demand assumptions and acceptability criteria to avoid unintended performance. Therefore, procedures incorporating inelastic analysis can reduce the uncertainty and conservatism.

This approach is also known as "pushover" analysis. A pattern of forces is applied to a structural model that includes non-linear properties (such as steel yield), and the total force is plotted against a reference displacement to define a capacity curve. This can then be combined with a demand curve (typically in the form of an acceleration-displacement response spectrum (ADRS)). This essentially reduces the problem to a single degree of freedom (SDOF) system.

Nonlinear static procedures use equivalent SDOF structural models and represent seismic ground motion with response spectra. Story drifts and component actions are related subsequently to the global demand parameter by the pushover or capacity curves that are the basis of the non-linear static procedures.

Nonlinear Dynamic Analysis

Nonlinear dynamic analysis utilizes the combination of ground motion records with a detailed structural model, therefore is capable of producing results with relatively low uncertainty. In nonlinear dynamic analyses, the detailed structural model subjected to a ground-motion record produces estimates of component deformations for each degree of freedom in the model and the modal responses are combined using schemes such as the square-root-sum-of-squares.

In non-linear dynamic analysis, the non-linear properties of the structure are considered as part of a time domain analysis. This approach is the most rigorous, and is required by some building codes for buildings of unusual configuration or of special importance. However, the calculated response can be very sensitive to the characteristics of the individual ground motion used as seismic input; therefore, several analyses are required using different ground motion records to achieve a reliable estimation of the probabilistic distribution of structural response. Since the properties of the seismic response depend on the intensity, or severity, of the seismic shaking, a comprehensive assessment calls for numerous nonlinear dynamic analyses at various levels of intensity to represent different possible earthquake scenarios. This has led to the emergence of methods like the Incremental Dynamic Analysis.

Earthquake Simulation

Earthquake simulation applies a real or simulated vibrational input to a structure that possesses the essential features of a real seismic event. Earthquake simulations are generally performed to study the effects of earthquakes on man-made engineered structures, or on natural features which may present a hazard during an earthquake.

Dynamic experiments on building and non-building structures may be physical – as with shake-table testing – or virtual (based on computer simulation). In all cases, to verify a structure's expected seismic performance, researchers prefer to deal with so called 'real time-histories' though the last cannot be 'real' for a hypothetical earthquake specified by either a building code or by some particular research requirements.

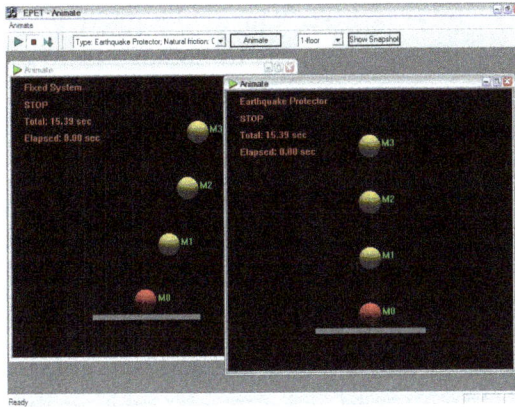

PrintScreen images of concurrent computer models animation

Shake-table Testing

Studying a building's response to an earthquake is performed by putting a model of the structure on a shake-table that simulates the seismic loading. The earliest such experiments were performed more than a century ago.

Computational Approaches

Another way is to evaluate the earthquake performance analytically. The very first earthquake simulations were performed by statically applying some *horizontal inertia forces*, based on scaled peak ground accelerations, to a mathematical model of a building. With the further development of computational technologies, static approaches began to give way to dynamic ones.

Traditionally, numerical simulation and physical tests have been uncoupled and performed separately. So-called *hybrid testing* systems employ rapid, parallel analyses using both physical and computational tests.

Earthquake Resistant Structures

Earthquake-proof and massive pyramid El Castillo, Chichen Itza.

Earthquake-resistant structures are structures designed to withstand earthquakes. While no structure can be entirely immune to damage from earthquakes, the goal of earthquake-resistant construction is to erect structures that fare better during seismic activity than their conventional counterparts.

According to building codes, earthquake-resistant structures are intended to withstand the largest earthquake of a certain probability that is likely to occur at their location. This means the loss of life should be minimized by preventing collapse of the buildings for rare earthquakes while the loss of functionality should be limited for more frequent ones.

To combat earthquake destruction, the only method available to ancient architects was to build their landmark structures to last, often by making them excessively stiff and strong, like the El Castillo pyramid at Chichen Itza.

Currently, there are several design philosophies in earthquake engineering, making use of experimental results, computer simulations and observations from past earthquakes to offer the required performance for the seismic threat at the site of interest. These range from appropriately sizing the structure to be strong and ductile enough to survive the shaking with an acceptable damage, to equipping it with base isolation or using structural vibration control technologies to minimize any forces and deformations. While the former is the method typically applied in most earthquake-resistant structures, important facilities, landmarks and cultural heritage buildings use the more advanced (and expensive) techniques of isolation or control to survive strong shaking with minimal damage. Examples of such applications are the Cathedral of Our Lady of the Angels and the Acropolis Museum.

Trends and Projects

Some of the new trends and/or projects in the field of earthquake engineering structures are presented.

Building Materials

Based on experience in earthquakes in Eastern European and in Central Asian countries where precast concrete has been widely used as construction material, it can be concluded that their seismic performance has been fairly satisfactory. Based on studies in New Zealand, relating to Christchurch earthquakes, precast concrete designed and installed in accordance with modern codes performed well. According to the Earthquake Engineering Research Institute, precast panel buildings had good durability during the earthquake in Armenia, compared to precast frame-panels.

Earthquake Shelter

One Japanese construction company has developed a six-foot cubical shelter, presented as an alternative to earthquake-proofing an entire building.

Concurrent Shake-table Testing

Concurrent shake-table testing of two or more building models is a vivid, persuasive and effective way to validate earthquake engineering solutions experimentally.

Thus, two wooden houses built before adoption of the 1981 Japanese Building Code were moved to E-Defense for testing. The left house was reinforced to enhance its seismic resistance, while the other one was not. These two models were set on E-Defense platform and tested simultaneously .

Combined Vibration Control Solution

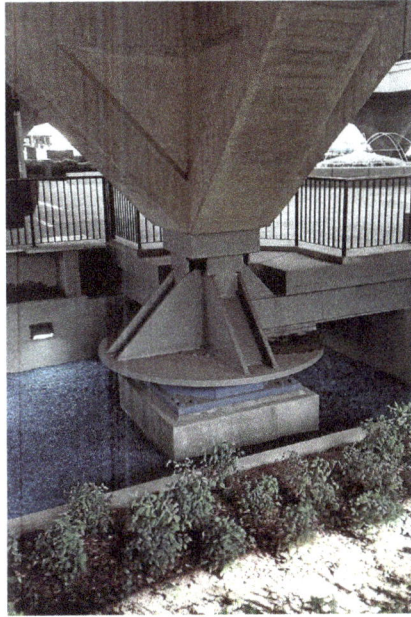

Close-up of abutment of seismically retrofitted Municipal Services Building in Glendale, CA

Seismically retrofitted Municipal Services Building in Glendale, CA

Designed by architect Merrill W. Baird of Glendale, working in collaboration with A. C. Martin Architects of Los Angeles, the Municipal Services Building at 633 East Broadway, Glendale was completed in 1966 . Prominently sited at the corner of East Broadway and Glendale Avenue, this civic building serves as a heraldic element of Glendale's civic center.

In October 2004 Architectural Resources Group (ARG) was contracted by Nabih Youssef & Associates, Structural Engineers, to provide services regarding a historic resource assessment of the building due to a proposed seismic retrofit.

In 2008, the Municipal Services Building of the City of Glendale, California was seismically retrofitted using an innovative combined vibration control solution: the existing elevated building foundation of the building was put on high damping rubber bearings.

Steel Plate Walls System

Coupled steel plate shear walls, Seattle

The Ritz-Carlton/JW Marriott hotel building engaging the advanced steel plate shear walls system, LA

A steel plate shear wall (SPSW) consists of steel infill plates bounded by a column-beam system. When such infill plates occupy each level within a framed bay of a structure, they constitute a SPSW system. Whereas most earthquake resistant construction methods are adapted from older systems, SPSW was invented entirely to withstand seismic activity.

SPSW behavior is analogous to a vertical plate girder cantilevered from its base. Similar to plate girders, the SPSW system optimizes component performance by taking advantage of the post-buckling behavior of the steel infill panels.

The Ritz-Carlton/JW Marriott hotel building, a part of the LA Live development in Los Angeles, California, is the first building in Los Angeles that uses an advanced steel plate shear wall system to resist the lateral loads of strong earthquakes and winds.

Kashiwazaki-Kariwa Nuclear Power Plant is Partially Upgraded

The Kashiwazaki-Kariwa Nuclear Power Plant, the largest nuclear generating station in the world by net electrical power rating, happened to be near the epicenter of the strongest M_w 6.6 July 2007 Chūetsu offshore earthquake. This initiated an extended shutdown for structural inspection which indicated that a greater earthquake-proofing was needed before operation could be resumed.

On May 9, 2009, one unit (Unit 7) was restarted, after the seismic upgrades. The test run had to continue for 50 days. The plant had been completely shut down for almost 22 months following the earthquake.

Seismic Test of Seven-Story Building

A destructive earthquake struck a lone, wooden condominium in Japan . The experiment was webcast live on July 14, 2009 to yield insight on how to make wooden structures stronger and better able to withstand major earthquakes .

The Miki shake at the Hyogo Earthquake Engineering Research Center is the capstone experiment of the four-year NEESWood project, which receives its primary support from the U.S. National Science Foundation Network for Earthquake Engineering Simulation (NEES) Program.

"NEESWood aims to develop a new seismic design philosophy that will provide the necessary mechanisms to safely increase the height of wood-frame structures in active seismic zones of the United States, as well as mitigate earthquake damage to low-rise wood-frame structures," said Rosowsky, Department of Civil Engineering at Texas A&M University. This philosophy is based on the application of seismic damping systems for wooden buildings. The systems, which can be installed inside the walls of most wooden buildings, include strong metal frame, bracing and dampers filled with viscous fluid.

Superframe RC Earthquake Proof Structure

The proposed system is composed of core walls, hat beams incorporated into the top level, outer columns and viscous dampers vertically installed between the tips of the hat beams and the outer columns. During an earthquake, the hat beams and outer columns act as outriggers and reduce the overturning moment in the core, and the installed dampers also reduce the moment and the lateral deflection of the structure. This innovative system can eliminate inner beams and inner columns on each floor, and thereby provide buildings with column-free floor space even in highly seismic regions.

Vibration Control

In earthquake engineering, vibration control is a set of technical means aimed to mitigate seismic impacts in building and non-building structures.

All seismic vibration control devices may be classified as *passive, active* or *hybrid* where:

- *passive control devices* have no feedback capability between them, structural elements and the ground;

- *active control devices* incorporate real-time recording instrumentation on the ground integrated with earthquake input processing equipment and actuators within the structure;

- *hybrid control devices* have combined features of active and passive control systems.

When ground seismic waves reach up and start to penetrate a base of a building, their energy flow density, due to reflections, reduces dramatically: usually, up to 90%. However, the remaining portions of the incident waves during a major earthquake still bear a huge devastating potential.

After the seismic waves enter a superstructure, there is a number of ways to control them in order to sooth their damaging effect and improve the building's seismic performance, for instance:

- to dissipate the wave energy inside a superstructure with properly engineered dampers;

- to disperse the wave energy between a wider range of frequencies;

- to absorb the resonant portions of the whole wave frequencies band with the help of so-called *mass dampers* .

Devices of the last kind, abbreviated correspondingly as TMD for the tuned (*passive*), as AMD for the *active*, and as HMD for the *hybrid mass dampers*, have been studied and installed in high-rise buildings, predominantly in Japan, for a quarter of a century .

Base-isolated San Francisco City Hall after seismic retrofit

However, there is quite another approach: partial suppression of the seismic energy flow into the superstructure known as *seismic* or *base isolation* which has been implemented in a number of historical buildings all over the world and remains in the focus of earthquake engineering research for years.

For this, some pads are inserted into all major load-carrying elements in the base of the building which should substantially decouple a superstructure from its substructure resting on a shaking ground. It also requires creating a rigidity diaphragm and a moat around the building, as well as making provisions against overturning and P-delta effect.

In refineries or plants snubbers are often used for vibration control. Snubbers come in two different variations: hydraulic snubber and a mechanical snubber.

- Hydraulic snubbers are used on piping systems when restrained thermal movement is allowed.

- Mechanical snubbers operate on the standards of restricting acceleration of any pipe movements to a threshold of 0.2 g's, which is the maximum acceleration that the snubber will permit the piping.

Superstructure

A superstructure is an upward extension of an existing structure above a baseline. This term is applied to various kinds of physical structures such as buildings, bridges, or ships having the degree of freedom zero (in the terms of theory of machines). The word "superstructure" is a combination of the Latin prefix, *super* (meaning *above, in addition*) with the Latin stem word, *structure* (meaning *to build* or *to heap up*).

In order to improve the response during earthquakes of buildings and bridges, the superstructure might be separated from its foundation by various civil engineering mechanisms or machinery. All together, these implement the system of earthquake protection called base isolation.

The superstructure of this cargo ship is in the back and includes a lifeboat.

The cruiseferry *Mega Smeralda*. The blue and white part of the ship is the superstructure and the yellow part of the ship is the hull.

Aboard Ships and Large Boats

As stated above, superstructure consists of the parts of the ship or a boat, including sailboats, fishing boats, passenger ships, and submarines, that project above her main deck. This does not usually include its masts or any armament turrets. Note that in modern times, turrets do not always carry naval artillery, but they can also carry missile launchers and/or antisubmarine warfare weapons.

The size of a watercraft's superstructure can have many implications in the performance of ships and boats, since these structures can alter their structural rigidity, their displacements, or both. These can be detrimental to any vessel's performance if they are taken into consideration incorrectly.

The height and the weight of superstructure on board a ship or a boat also affects the amount of freeboard that such a vessel requires along its sides, down to her waterline. In broad terms, the more and heavier superstructure that a ship possesses (as a fraction of her length), the less the freeboard that is needed.

Bridges

On a bridge, the portion of the structure that is the span and directly receives the live load is referred to as the superstructure. In contrast, the abutment, piers, and other support structures are called the substructure.

References

- Bozorgnia, Yousef; Bertero, Vitelmo V. (2004). Earthquake Engineering: From Engineering Seismology to Performance-Based Engineering. CRC Press. ISBN 978-0-8493-1439-1.

- Chu, S.Y.; Soong, T.T.; Reinhorn, A.M. (2005). Active, Hybrid and Semi-Active Structural Control. John Wiley & Sons. ISBN 0-470-01352-4.

- Arnold, Christopher; Reitherman, Robert (1982). Building Configuration & Seismic Design. A Wiley-Interscience Publication. ISBN 0-471-86138-3.

- Reitherman, Robert (2012). Earthquakes and Engineers: An International History. Reston, VA: ASCE Press. pp. 394–395. ISBN 9780784410714.

- EERI Endowment Subcommittee (May 2000). Financial Management of Earthquake Risk. EERI Publication. ISBN 0-943198-21-6.

- Craig Taylor; Erik VanMarcke, eds. (2002). Acceptable Risk Processes: Lifeline and Natural Hazards. Reston, VA: ASCE, TCLEE. ISBN 9780784406236.

- Lindeburg, Michael R.; Baradar, Majid (2001). Seismic Design of Building Structures. Professional Publications. ISBN 1-888577-52-5.

- Reitherman, Robert (2012). Earthquakes and Engineers: An International History. Reston, VA: ASCE Press. pp. 356–357. ISBN 9780784410714.

- A Survey on concepts of design and executing of Superframe RC Earthquake proof Structures" (2016) by Kiarash Khodabakhshi ISBN 9783668208704

- Chu, S.Y.; Soong, T.T.; Reinhorn, A.M. (2005). Active, Hybrid and Semi-Active Structural Control. John Wiley & Sons. ISBN 0-470-01352-4.

- "Bad construction cited in quake zone - World news - Asia-Pacific - China earthquake | NBC News". MSNBC. Retrieved 2013-07-28.

- "simulacion terremoto peru-huaraz - casas de adobe - YouTube". Nz.youtube.com. 2006-06-24. Retrieved 2013-07-28.

- "Shake table testing of adobe house (4A-S7 East) - YouTube". Nz.youtube.com. 2007-01-12. Retrieved 2013-07-28.

- "Strategy to Close Metsamor Plant Presented | Asbarez Armenian News". Asbarez.com. 1995-10-26. Retrieved 2012-07-31.

- "Building Technology + Seismic Isolation System - OKUMURA CORPORATION" (in Japanese). Okumuragumi.co.jp. Retrieved 2012-07-31.

- "CMMI - Funding - Hazard Mitigation and Structural Engineering - US National Science Foundation (NSF)". nsf.gov. Retrieved 2012-07-31.

Permissions

Index